A Primer on Auction Design, Management, and Strategy

A Primer on Auction Design, Management, and Strategy

David J. Salant

The MIT Press
Cambridge, Massachusetts
London, England

MIT Press books may be purchased at special quantity discounts for business or sales promotional use. For information, please email special_sales@mitpress.mit.edu.

This book was set in Palatino LT Std by Toppan Best-set Premedia Limited, Hong Kong. Printed and bound in the United States of America.

Library of Congress Cataloging-in-Publication Data

Salant, David J.
A primer on auction design, management, and strategy / David J. Salant.
pages cm
Includes bibliographical references and index.
ISBN 978-0-262-02826-4 (hardcover : alk. paper)
1. Auctions. 2. Game theory. I. Title.
HF5476.S24 2014
658.8'77–dc23

2014013241

10 9 8 7 6 5 4 3 2 1

Contents

Preface

My work on auctions started in 1994 when I was a Senior Member of Technical Staff at GTE Laboratories (now Verizon Laboratories) and was asked to make a one-hour presentation on auctions to some executives involved in the planning for the first US spectrum auctions. That one-hour talk evolved into twenty years of auction consulting assignments.

This work evolved still further in mid-2007, when Patrick Rey asked if I might be interested in visiting the Toulouse School of Economics (TSE) and teaching a course on auctions. I accepted Patrick's offer, and this book is a compilation of the lecture notes I put together in teaching this class for each of the past six years at TSE. This work has benefited enormously from the hospitality I received at Toulouse and the feedback of my students.

The TSE has been the ideal place for me to develop the material contained in this Primer. My goal has been to provide a rudimentary introduction to auctions to someone who is potentially considering running an auction or bidding in one. This does require some familiarity with basic concepts of game theory and some results in auction theory.

There are a number of existing books that provide excellent coverage of auction theory—notably Milgrom's *Putting Auction Theory to Work*, Klemperer's *Auctions: Theory and Practice*, and Krishna's *Auction Theory*. I felt that there was still a large gap for the practitioner. Notably, it seems, at least based on my experience, that it is all too easy to misinterpret or misapply some of the theory, or all too difficult to apply that theory appropriately. These other works, while providing very valuable structures and case studies, leave a lot of gaps, requiring an auctioneer or a bidder to fully understand the limits of this theory in interpreting and in applying this theory. I have tried, as much as possible to help bridge these gaps. This Primer doesn't cover as much of the theory, or in as much detail. However, I try to explain in some detail how the theory has been or can be applied to the decision problems that auction participants face. To this end, I have included game theory at a level needed to use and avoid abusing the main basic results from auction theory, and at a level for the practitioner to be able to critically evaluate prior auction experience.

A step-by-step manual to run or bid in an auction was never my intent in writing this book. It was instead my intent to explain how to go about planning to run or bid in an auction. As the book grew in part out of a course I have taught at the TSE, it is probably more of a textbook than I had originally intended. It can probably serve as a textbook on auctions for graduate students in economics and other areas, as well as upper level undergraduates. And perhaps a good text is the best form of Primer.

This book has also benefited enormously from my work on consulting projects. I have had the privilege of being able to advise bidders and auctioneers in dozens of auctions over nearly twenty years. This experience has been a source of many of the examples used and influenced the material in the book. Indeed some assignments have motivated new research—for example, the "chopstick" paper of Rosenthal and Szentes (2003) was motivated by the Dutch 2G auction. And much of the section on sequential auctions was motivated by my work on the New Jersey BGS auction (and published in Loxley and Salant 2004).The experience is also a great source of examples of how auctions can fail, because of design flaws and possible bidding miscalculations, and not just a lack of interest from bidders.

One of the goals of the book is to explain the decision problems facing bidders; so while the revenue equivalence theorem states that a bidder should expect to pay the same amount in a first-price and second-price sealed-bid auction, the decision problem facing bidders is not at all the same. Bidders should bid values in a second-price auction for a single object. In contrast, in first-price auctions bidders need to calculate an optimal bidding strategy—which is not all that simple and requires some assumptions about rivals.

I have benefited from comments of Elmar Wolfstetter who has provided comments on the entire manuscript. My work on chapter 10 benefited a great deal from a consulting project with Andy Skrzypacz and Jon Levin. Paul Klemperer read through the manuscript and provided many helpful comments. In fact this Primer owes an enormous intellectual debt to Klemperer's work, and especially his *Auctions: Theory and Practice*. I have also had the truly distinct privilege to work with and learn from Paul Milgrom. Milgrom has made enormous efforts to develop auction designs that improve market efficiency. His uncanny ability to express the most complex ideas in easy to grasp concrete terms has been pivotal in getting new types of auctions introduced for new types of transactions. Over time, my appreciation of his skill and the effort he makes to find the best way to explain a difficult concept has grown.

The past twenty years have been a golden age for game theory. When I finished graduate school in 1979, game theory was just starting to become incorporated into standard economics curriculum. Until the first Federal Communications Commission (FCC) auctions in 1994, game theorists, and even more generally mathematical economics, had little impact outside of academe. The FCC auctions changed all that. For the first time, mathematical economics and game theory had a direct value to very real problems. Since the first FCC auctions, auction theorists have been involved in the design of auctions in a great many sectors—from airwaves to milk powder, gas and electricity to pollution rights. However, there is still much to be done I hope this Primer will expand the scope of the application of game theory to address practical problems.

Still there are a few significant obstacles to the application of concepts from modern auction theory to practical problems. One is that there are often parties that have a vested interest in markets and who would oppose new auctions that would improve market efficiency. For example, an auction can reduce the value of brokers or totally eliminate the need for brokers. Thus brokers who have the best access to the information needed to design and set up an auction may also very much oppose it. Another main impediment to the introduction of new types of auctions is normal business inertia. If an existing system for trade seems to work reasonably well, then buyers and sellers may not see a need to try to improve it. This is partly due to the costs of learning a new system. And inertia can set in even where auctions have had some success. For example, I introduced the simultaneous descending clock auction (SDCA), a variant of the simultaneous multi-round auction (SMRA) for energy procurement. These auctions typically take several days to complete—but given the experience with the SMRA and SDCA, these auctions can be completed within hours and not days. For the larger energy procurement auctions, the transaction costs of bidders are significantly higher when their offers have to be held open several days. But the electric utilities that run the auctions and the regulatory agencies that oversee have no desire to fix an auction that is not broken. The best opportunities for introducing new auction formats are often immediately after there has been some sort of problem, such as an energy crisis, or one with mortgage-backed securities.

Last, I wish to express my appreciation to my family. My wife Deborah, and my children Jacob, Rebecca, Ilana, and Max have had to put up with my long absences and constant traveling during the time I was accumulating much of the experience that is the source of much of the material in this book.

1 Introduction

Summary

This is a Primer on auctions. It is intended, in part, to serve as a guide to auctions for the practitioner. This chapter explains what distinguishes auctions from other market transactions. It provides a brief summary of what is essential or helpful for the practitioner.

1.1 Goals of This Primer

Auctions are an increasingly common share of all market transactions. They take a great many forms. Often they greatly improve market liquidity, but they can totally fail to work. Perhaps more than half of all auctions result in some or all of what is being put for auction remaining unsold, and even when this does not occur, the transaction prices can be absurdly low or high, and far from equilibrium levels. This book explores reasons why auctions work or fail. It explains how auction design affects bidding decisions and how bidding decisions affect the outcome, sometimes quite dramatically.

Auctions are highly structured market transactions primarily used in thin markets (markets with few participants and infrequent transactions). In such markets the standard supply-and-demand models of competitive market forces usually cannot be relied on to explain the outcomes, since in most auctions just a few bidders can have a large influence on the outcome. An auction is perhaps better characterized as a bargaining process. Moreover in auctions, unlike most other markets, offers and counteroffers are typically made within a structure defined by a set of rigid and comprehensive rules. This book provides a fairly complete range of auction structures and discusses how they can affect auction outcomes.

This book is intended to serve both as an introductory text on auctions as well as a guide for practitioners, that is, those interested in managing or bidding in auctions. Auctions are highly structured negotiations with a defined set of rules for offers, counteroffers, price determination, and allocations. These rigid structures mean that auctions can be analyzed by mathematical models more accurately and completely than can most other types of market transactions.

This book provides a guide for modeling, analyzing, and predicting the outcomes of auctions. To be useful and effective, such a guide must cover some essential elements of auction theory. In practice, auction theory will rarely provide an exact and prescriptive model that can be applied directly. Nevertheless, auction theory does provide both insights and specific results that are of direct value to the practitioner.

While a primer or practitioner's guide would ideally provide a complete checklist for what to do to design, set up, and manage an auction, or for bidders to decide how to bid, that scope is too broad to be practical. Rather, the focus here is on principles, tools, and examples that can be used to analyze auctions. This first requires some way to categorize different types of auctions. In this categorization, I explain how the auction design and strategy should depend on the type of auction. Second, I explain the main results from game theory and the theory of auctions in order to provide practical frameworks for analyzing auctions and bid strategies. Third, I review auction experience, mainly from actual auctions but also from experiments. This part of the book illustrates how auction theory can be applied to auction design and strategy decisions.

Because of the increased use of auctions in new settings, auctions have become a more common subject of economic research and writing. Because this Primer is intended as a guide for someone interested in designing an auction and managing it or for someone having to bid, its review of research on game theory and auction theory will stress the linkage between theory and application. Some game theory is crucial for the principles of sound auction design and for properly assessing bidding strategy.

This book largely avoids using formulaic and specific task plans or checklists because most auction design and bid strategy problems are too complex to lend themselves to exact formulas. However, one of the main points of this book is to show how to apply game-theoretic modeling tools to practical auction design and strategy decisions. Practical difficulties are explained in some detail. Reality never exactly fits any theoretical model, but very simple theoretical models can still be very useful.

Explanations of how auction strategy and auction design can matter in one specific case may have only limited applicability to seemingly very similar specific cases. Similarly the main results of auction theory may have only small direct practical value. What is often most useful are simple mathematical models and a few basic principles for developing them, as well as guidelines to help identify limitations of such models. *Simulations,* in which computer algorithms substitute for bidders, and scaled-down *experiments,* in which human subjects play the role of bidders in repeated trials, can also be useful tools for assessing auction design or bid strategy plans. Much of this book's exposition is devoted to explaining how to develop simple and useful models of auctions.

One additional objective in this book is to explain how both auction design and bid strategy can have a significant effect on prices and allocations. Auctions are a form of imperfectly competitive markets, in which both sides have limited information. More specifically, auction design can have a significant effect on revenues in a *forward* auction, that is, an auction to sell one or more lots, and on costs in a *reverse,* or procurement, auction. I provide many examples in which identical items have sold for significantly different prices in the same auction and at the same time. An analysis of how the auction design affects bidder incentives can usually go a long way in explaining such price anomalies. Price differences are even more common when identical objects are auctioned one at a time and in other types of sequential transactions.

I start with a basic exposition of game theory and with a few of the main results from auction theory. To be useful, a model of an auction must be consistent and must quantitatively represent the incentives and decision-making of the bidders. Mostly this assumes rational bidders. However, behavioral and cognitive limitations of bidders are discussed.

This Primer should not be considered as a comprehensive compendium of modern auction theory. Many topics are omitted—such as two-sided auctions or econometric analysis of auctions[1]—and there is relatively limited discussion of some more technical topics, such as mechanism design. The intent here is to provide an introduction and a starting point for the reader to either learn something about auctions or interested in auction design or strategy.[2] This Primer still should also be of interest to researchers. There are many practical issues in auction design and strategy that have yet to be studied by researchers. Some of the material in this book is new, largely motivated by a need to extend and expand the existing literature in order to address some design issues that have arisen in recent auctions.

1.2 What Are Auctions?

This section briefly describes what auctions are, and the types of auctions studied in this book. I also provide a brief introduction as to why auctions are used in preference to other market mechanisms.

To start, it is essential to say what constitutes an auction, or at least what will be included in this book. What distinguishes an auction from other types of transactions is not always clearly specified. I will use the term primarily to describe all one-to-many bidding processes: forward auctions, with one seller and many buyers, or reverse auctions, with one buyer and many sellers. At times the term *auction* is extended to situations in which there are a few buyers or a few sellers, namely few-to-many transactions. But in these cases the few will be assumed to be acting in concert in organizing the auction. When there are two sellers, acting totally independently, there are really two independent auctions. In contrast, when the two sellers agree on schedules, starting prices, or other elements of the bidding process, then the two sellers are really acting as a single entity in determining the rules of the auction.

The term auction will be restricted here to what are often called *sealed-bid tenders*. More specifically, one-shot bidding processes, both price-only and multi-attribute ones, are referred to here as sealed-bid auctions. Most of the focus is on price-only or other forms of single-dimensional bidding.

More formally, an auction is defined by the following:

1. *Bidding format rules* These rules govern the form of bids. Bids can be price only, multi-attribute, price and quantity, or quantity only.
2. *Bidding process rules* Included are closing rules, which determine the timing of the auction; specification of the information provided to bidders during the auction; rules about the processes for bid improvements or counterbids; and rules specifying the auction-closing conditions.

3. *Price and allocation rules* These rules determine final prices and the amounts won by each winning bidder. Auctions can be for single items or multiple items. Traditionally multi-lot auctions have been conducted as separate auctions, sometimes sequentially and sometimes simultaneously. Modern auction designs have been developed to facilitate bidder arbitrage for similar items and/or to allow bidders to submit offers for different packages of items.

An auction is one type of market mechanism. More specifically, an auction is an allocation mechanism that mostly uses price criteria for determining *allocations*—that is, which parties win which objects or contracts—and the prices paid. Auctions are more centralized than most other market mechanisms. In most markets, other than *exchanges*, buy and/or sell offers are not all made in one place or submitted through an auctioneer. An auction, however, is not the only type of centralized market mechanism.

An auction is also a price discovery mechanism. Auctions provide a means for the entity conducting the auction—the auctioneer, or the auction *originator*, to collect information from bidders so as to determine prices and allocations. The fact that the information bidders provide can affect their prices means that it can also affect their incentives, as is explained in more detail below.

An exchange (e.g., a stock, commodity, or financial exchange) is at times considered to be a form of auction. Exchanges differ from the auctions considered here in three ways. First, exchanges involve two-sided or many-to-many transactions. Second, transactions on an exchange normally run continuously, whereas auctions are periodic or episodic. Third, exchanges usually have an intermediary setting the price who is neither a net buyer nor a net seller on average. The intermediary (or specialist or trading desk) will have a book of buy and sell orders and adjust prices up or down depending on the relative volumes of such orders. In contrast, in auctions the originator will usually have an interest in getting the best price (i.e., the highest price in a forward auction, and the lowest price in a reverse auction). Auctions can, however, have independent entities to manage the bidding process.[3]

1.2.1 Why Auctions?

Auctions are not used for most transactions. The question arises why auctions are used at all. When the transaction frequency is high, transactions occur at physical or virtual stores, or in exchanges. When it is low, but the volume or monetary value is high, an auction is most useful. When the transaction frequency is low, it will not be worthwhile to incur the cost of maintaining a trading site, or store. While the organization of an auction may involve some costs, when the trading events are infrequent, it can be cost effective for auctions to be used.

When the volume or the value is also low, bilateral negotiations and the solicitation of offers through requests for offers (RFOs) or requests for proposals (RFPs) are often most efficient. When the transaction frequency is low, the cost of maintaining a store or exchange is high relative to the transaction value. When the transaction value is low, the cost of organizing and managing an auction can be prohibitive. Thus auctions are best for periodic or episodic high-volume and high-value transactions.

1.2.2 Types of Auctions

Auctions have been around for centuries—at least since the time of Herodotus in 500 BCE. Auctions can take many forms and can involve many types of goods and services. Two main ways in which auctions differ are in the auction format, or rules structure, and the frequency and volume of the transactions. Until recently, most auctions were single-object auctions or sequences of such auctions.

Auction Formats Traditionally most price-only auctions used some variant on one of four basic forms:

1. English, ascending price
2. Dutch, descending price
3. First-price sealed bid
4. Second-price sealed bid

Some auctions allow bids to include multiple components or attributes, and not just price. In multi-attribute bidding processes explicit weights or subjective evaluations are used to decide winners.

The most common form of open auction is still the open, oral-outcry *English* auction. In a standard English auction, the auction manager or auctioneer announces prices—an increasing sequence of prices in a forward auction, and a decreasing sequence in procurement auctions—and bidders indicate whether they are willing to accept the announced price. Bidders need not indicate a willingness to accept each announced price in order to remain eligible, and once one bidder indicates acceptance of the most recent price, the auctioneer will go to the next increment or decrement. Bidders can often, if they desire, shout out (inject) jump bids. The auction ends when no one is willing to improve on the most recently announced price.

The most common form of *online* auction is still the English auction, or simple variants thereof. Instead of an auctioneer calling out prices, though, the host sets a starting price, and bidders can submit bids at any time during an open bidding window. Each new bid is posted, and the auction host will raise the minimum for subsequent bids, similarly to the manner in which a live auctioneer raises price in an English auction.

Descending-price forward auctions have been used since Babylonian times.[4] The name "Dutch auction" to describe these types of auctions probably arose from their use in selling tulips in the Netherlands, which dates back at least to the seventeenth century.[5] In a (forward) Dutch auction, the auctioneer starts with a high price, and gradually lowers the price until a bidder is willing to accept that price. Notice that for a reverse auction, that is, an auction in which bidders are sellers and the auctioneer is a buyer, a declining-price auction would elicit bids from all bidders at a high initial price. So, strategically, a declining-price forward (Dutch) auction, that is, an auction in which bidders are buyers, is strategically quite different from a declining-price reverse auction. Here, the term "Dutch" is reserved for declining-price forward auctions. The bid strategy is clearly different in a Dutch and an English auction. In a Dutch auction, a bidder has to guess when to bid. In an English auction, there is no such guessing. A bidder can stop when his or her value is reached.

Both first-price and second-price sealed-bid auctions, often called sealed-bid tender auctions, award to the high bidder the object for sale in a forward auction, and to the low bidder the contract to supply a product or service in a sealed-bid reverse auction. In a first-price auction, the winning bidder pays the bid amount in a forward auction, or is paid the bid amount in a reverse auction. In a second-price auction, the winning bidder pays, or is paid, the amount offered by the best losing bid.

Transaction Frequency and Volume Auctions also differ in how many units are transacted and how often. An auction for a single object is a one-time event, and the strategic considerations are quite different than they are in multi-object auctions and repeated auctions. The theory and experience for one-shot auctions is a useful reference point for the subsequent analysis of multi-object auctions. Auctions can be one-time, or infrequent, events for a large number of objects. The major spectrum auctions in the United States, Europe, and elsewhere are examples of such auctions. Such auctions have a *matching* role in determining which objects go to which bidders, as well as the usual allocation role in determining winners and losers. Bidders can also interact in the auction to try to more efficiently sort out who wins what, and at what prices. It is important that the auction rules allow efficient arbitrage, as otherwise both the auctioneer and the bidders can lose.

Auctions can also be recurring, such as the monthly capacity auctions conducted by many electricity transmission operators to ensure adequate generation availability. Many commodities—dairy products and wine, for example—are auctioned periodically. Mostly the same bidders compete in each such auction, and for mostly the same types of products. These periodic auctions are commonly multi-unit, and often multi-object. As such, they have many of the same strategic features as one-shot, multi-object auctions. However, the repetition allows bidders to communicate and interact over time across auctions. This can further affect bidding behavior and the outcomes of the auctions.

Auctions can be categorized in other ways. *Forward* auctions—that is, auctions in which there is one seller and several buyers—are perhaps the most common. But *reverse* auctions, in which there is one buyer and several sellers, are also fairly common. The modeling and analysis of forward and reverse auctions tend to be very similar, save for the fact that in reverse auction, the auctioneer needs to ensure post-auction delivery, rather than just a monetary payment in a forward auction.

1.3 A New Age of Auctions

Auctions have changed. They come in a great many new forms. A recent search at the US Patent Office website produced over 2,000 hits for patents issued that include the term "auction." While not every patent is for a different auction design, the number of innovations is still large. There are two main reasons for these changes, both in some sense owing something to John von Neumann (1903–1957), a mathematician who made pioneering contributions to both computer science and game theory.

One reason is the developments in game theory, a field that barely existed until the 1960s. Game theory is crucial, or should be crucial, to auction design. Game theory provides mathematical tools and techniques to analyze alternative auction designs. While game theory, and game theorists, are not always directly involved in auction design, their role has increased over the past ten to fifteen years, both in government agencies (e.g., the US Treasury, other national treasuries, and other governmental agencies in a number of sectors including telecommunications, energy, and natural resources) and in the private sector (where firms, e.g., Google, Microsoft, Yahoo, HP, and IBM, have hired leading game theorists to help improve their pricing tools: remember that an auction is a price discovery mechanism).

The other area in which von Neumann made pioneering contributions was in computer science. Without computers and the Internet, the use of auctions might still be where it was centuries ago. To say that auctions are a price discovery mechanism means bids provide the auction originator information about valuations, which are then used to determine prices and allocations. However, bidders will have incentives to withhold information if the information will affect prices paid. Iterative bidding, and other competitive processes, can mitigate bidders' incentives for strategic withholding. These iterative and competitive processes require efficient data exchange and processing facilities—facilities that would not otherwise be available without computers. Moreover the use of the Internet greatly expands the pool of potential auction participants and reduces participation costs. These factors tend to improve auction results, and make auctions feasible for transactions that could never be realized absent computers and electronic communication networks.

1.3.1 New Types of Auctions

Over the past twenty years there has been a proliferation of different auction types. This section does not catalog all of them. Instead, it briefly explains a few of the ways in which auction design has changed, due in part to changes in electronic communication technology, and in part to developments in game theory. The former allows more information to be transmitted and processed more quickly. The latter provides better incentives, and allows auctions to be introduced in areas where they may never have been possible before.

There have been significant advances in multi-object auctions. A large class of simultaneous auctions for buying or selling a number of substitutes or complements at one time includes the *simultaneous multi-round* (SMR) *auction*, the *simultaneous ascending auction* (SAA), the *simultaneous ascending and descending clock auctions*,[6] and package bid auctions, the Anglo-Dutch hybrid auction, simultaneous auctions with intra-round bidding, and many other designs. They have been introduced building off developments in game theory for auction design on the one hand, and in computer and communication technology on the other. In addition a few types of package bid auctions have been introduced, including the SMR auction with hierarchical package bidding used for the $19 billion auction of 700 MHz licenses conducted by the US Federal Communications Commission in 2008.

Even simple auction designs have changed. As noted above, online auctions typically employ a variant of the standard English auction in which a bidding window is open for a fixed duration, and bidders can enter offers as long as their offer exceeds the previous high bid by a minimum amount and the bidding window is open. This type of auction has been referred to as a "Yankee" auction. It has many variants, including one that extends the bidding window any time a new bid is entered. eBay has a "buy-it-now" feature that allows a bidder to enter a bid and close the auction.

Computers make possible package, or combinatorial, bidding. A package bid is an all-or-nothing bid on a combination of items. Package bidding is intended to address the *exposure* problem: a bidder having a value for a combination or package of items, but having a zero or sufficiently low value for the individual items, will be reluctant to submit bids on the individual items and risk winning them, unless that bidder is fairly certain it will win all the items in the package. Package bidding eliminates this exposure risk. However, package bidding adds a great deal of computational complexity because of the large number of possible combinations. When there are N items available, there will be 2^N possible subsets.[7] This number will exceed 1,000 with only 10 items available.

Many new designs address the form of bids and the way bids are submitted to the originator.

1.3.2 Auctions Replacing Regulation

One area in which auctions have become an increasingly accepted transaction mechanism is in regulation, especially in the energy and telecommunications sectors. New Zealand and the United States were the first countries to introduce auctions for allocating spectrum rights.[8] Prior to auctions being used for allocating spectrum rights, a variety of allocation rules—including first-come, first-served; lotteries;

and comparative hearings—were used. Auctions have now been introduced for allocating spectrum rights in dozens of countries. Spectrum auctions have perhaps been the largest auctions in history; the three largest spectrum auctions to date alone generated over $100 billion in revenues—nearly $50 billion for the German 3G auction, nearly $35 billion for the UK 3G auction, and almost $19 billion for the US 700 MHz auction.

Not all spectrum auctions have been successful. The success of the New Zealand auctions is debatable in view of their evidently low revenues and inefficiencies.[9] The US FCC's auction 5 resulted in most of the largest bidders declaring bankruptcy shortly after the auction. These bidders tried to renegotiate lower prices, and a number of the large winners were successful in doing so. It took almost ten years and a Supreme Court decision to resolve the conflict between bankruptcy law and the auction rules (*FCC v. NextWave Personal Communications, Inc.*, 537 US 293, 302 (2003)).

Despite a few such problems, auctions appear to be a tool gaining increasing acceptance from regulatory agencies. To date the US Federal Communications Commission has conducted over fifty auctions, the UK, Australian, and Canadian telecommunication regulatory authorities have conducted over half a dozen auctions each, and telecommunication regulatory agencies in many other countries have used auctions for selling spectrum rights. Auctions therefore appear to be a permanent part of the regulatory process for managing spectrum in a great many countries. Spectrum auctions will be discussed in more detail in chapter 2.

Auctions are also increasingly common in the energy sector for electricity and gas. Simultaneous descending clock auctions are or have been used for energy procurement in a number of US states, starting with New Jersey and now including Montana, Illinois, Ohio, and California, and also in Spain and Italy. Other simultaneous auction formats are or have been used for selling entitlements for electricity in Belgium, France, Germany, Alberta (Canada), and Texas. Sealed-bid auctions are being used for energy procurement in Maryland, Delaware, Virginia, and the District of Columbia. Auctions are also being used for selling transmission or interconnector rights in the United States and Europe, and for capacity transactions in much of the United States. In the gas sector, various versions of the simultaneous multi-round clock auction have been conducted in Austria, Germany, and Hungary for capacity rights.

Not all the experience has been positive. California has had particularly troubling experiences—in 1993, and then during the summer and fall of 2000.[10] The failure of California Power Exchange, a set of daily and longer term auctions, is a well-documented example of how not to run an auction. The California QF auctions described below resulted in *negative* energy prices. And some Texas capacity entitlement auctions resulted in difficult-to-explain 50 percent or greater price differences for identical products in the same auction. Both of these auctions are described in more detail in chapters 9 and 10.

In both the energy and telecommunication sectors, participation of qualified bidders is essential for a successful auction. Bidders don't always show up in sufficient numbers to make auctions competitive. Some examples in telecommunications include the first US auctions for 700 MHz spectrum and wireless communications services in 2000 and in 1997. In the latter case the forecast revenues were about 100 times the actual auction revenues. In particular, the Congressional Budget Office had forecast a value of $1.8 billion, but the auction raised only $13.6 million.[11] Similarly the Ohio First Energy procurement auctions have failed over several years to attract adequate competition to allow First Energy to purchase any of its default service resources through the auction.

1.3.3 Auctions in the Private Sector

There may be more auction activity in the private sector than in the public sector, but much less information is publicly available about private sector auctions. Information about wholesale or business-to-business (B2B) transactions tends to be confidential. There has been an increasing reliance on auctions for both buying and selling in B2B dealings. Auctions are being used for everything from selling agricultural commodities to procuring professional services. A significant number of firms have arisen in the B2B sector, and many have folded or merged. Among the notable firms still around as of 2013 were Perfect Commerce (formerly Commerce One and Perfect), Ariba-Procuri (which includes what were formerly called Ariba, Procuri and Trading Dynamics), and Dovebid. Many of these online B2B auction firms have migrated into supply chain management, and many others have focused on one or two specific supply chains or "verticals," such as Hambricht for IPOs, ChemConnect for chemicals, and Nexant for energy.

For retail auctions, consumer-to-consumer (C2C) and business-to-consumer (B2C) eBay and Yahoo have the largest market shares. What is important for consumer auctions is participation, and so there has been some tendency toward increasing concentration on the consumer side. However, specialized sites such as those for travel and entertainment have been able to maintain some market presence. Consumer auctions tend to be less rigidly structured, be less promoted, and have lower participation than do B2B or government-run auctions. For this reason, less of what follows applies to these consumer auctions. What has also been notable is that many of the online auction operations have encountered difficulties. There are a few exceptions, notably the online auction site eBay; and for ads, Google and Yahoo, which rely heavily on an auction pricing mechanism to price position on their search pages.

1.4 Why Auction Design (and Management) Matters

What seems to be clear is that no one auction approach works all the time. Paul Klemperer states that auction design is a matter of the "horse for the courses" and not "one size fit all."[12] There are several reasons auction design matters.

First, not all auctions work equally well. There are many things that can go wrong. In some cases the possibility of misallocations caused by an auction design that puts bidders in a position where they are prone to make miscalculations and bad guesses tends not to matter as much when there are many competitors.

Second, in some cases, notably in common value or affiliated value auctions, vigorous competition can result in overbidding and ex post performance problems.[13]

Third, in multi-lot auctions, misallocations can occur for several reasons. A great deal of auction design work has been devoted to multi-product and multi-unit auctions.[14] One of the main concerns in multi-object auctions is the exposure problem, described above, where a bidder may have a low value for the individual items in a package on which it places a high value. Package bidding and simultaneous auction designs, including the SMR and the SMR with both hierarchical and nonhierarchical package bidding, were designed in part to address this concern. Auctions without package bidding are more likely to leave bidders unable to obtain efficient combinations of lots.

Package bidding raised other concerns, one of which is the *threshold problem*. When two (or more) bidders are seeking individual items for which a third bidder has placed a package bid, then the two bidders may need to find a way to coordinate their bids.

For example, suppose that there are three bidders, B1, B2, and B3, and two items, I and II. Suppose that B1 wants item I, B2 wants II, and B3 wants both. If there is not much competition for I or II alone, neither B1 nor B2 might have to offer very much to be the high bidder on those items. Now suppose that B3 has placed a bid of 8, which is posted, for both items together, and that B1 puts a value of 6 on item I and B2 puts a value of 6 on item II. If, absent B3, these two items would sell for 2 each, then B2 and B3 would each need to at least double their bids to win. Neither may want to do so without knowing how much the other might want to offer. This can result in B3 winning, when B1 and B2 should. The reverse can happen when B3 has the high combined value but cannot enter a package bid. This possibility has been a topic of active research,[15] and recently the UK government has introduced versions of *core-selecting* package auctions to address it. Chapters 9 and 10 presents some of the theory of SMR and package auctions. Chapters 9 and 10 also present experience with multi-object auctions in the telecommunications and energy sectors. Package bidding also can add significant complexity. The number of possible combinations can get unmanageably large very fast.

One additional reason auction design matters is that some auctions are better at solving coordination problems for bidders.

On the flip side, there are instances in which it is possible for bidders to coordinate so as to divide the market and limit competition.[16] As a response, regulatory authorities responsible for setting auction rules have limited the information reported to bidders during an auction. The better bidders are able to communicate, the easier is coordination, and at times such coordination can serve primarily to reduce competition in an auction. So the form of bids and the information reported to bidders can affect competition in an auction.

As much as design matters, how well an auction does also depends a great deal on implementation. It is important to emphasize that a primary requisite for any auction is well-prepared bidders. The rules, procedures, and management must facilitate bidder participation and not deter it. Too often the reverse is true, often because the originator wants to guarantee the best outcome from his or her perspective. By optimizing the bidding rules for the originator, participation is discouraged, which paradoxically results in bad outcomes for the originator. And a competitive auction with less than ideal rules can often result in more efficient outcomes than will a good auction design with low participation. What is required is to enhance prospects for participation. What measures need to be taken so that participation does not sacrifice post-auction performance is discussed in chapter 11.

However, auctions with multiple units of the same object do not always result in uniform prices. One of the great successes in auction design is the development of efficient auction designs for multi-unit auctions that are easy to run. Clock auctions are quite efficient, and fast, for multi-unit, single-product auctions. Clock auctions and SMR auctions also work quite well when the products are substitutes and there are no large bidders. As has been noted elsewhere,[17] however, clock and SMR auctions can be vulnerable to collusive or coordinated bidding. Klemperer designed the Anglo-Dutch hybrid to try to strike a balance between the benefits of an open SMR auction and the risks of collusion.

Many new auction designs involve improvements to the bid submission process or to accommodate online bidding. The time restrictions imposed in many online auctions, specifying a narrow bidding window, create incentives for bidders to wait until the last possible instant to bid—an activity called *sniping*. The clock auction eliminates this incentive, as the auction manager, rather than the bidder, raises the price. However, for reasons explained below, clock auctions are less practical for many consumer and B2B transactions. This has spawned a great deal of effort to adjust for sniping.

1.5 Outline of This Primer

This remainder of the Primer is divided into three parts: (1) theory; (2) auction practice—that is, design, organization, and management; and (3) experience.

The theory part starts with a very brief introduction to basic concepts from game theory. The game theory framework is an essential tool for analyzing auctions. I then explain a number of the more significant results from the auction theory literature, including the revenue, or payoff, equivalence theorem, the winner's curse, optimal auction design, and the theory of simultaneous and sequential auctions. The revenue equivalence theorem is worthy of specific mention as a key result that has widespread applicability. The analysis of simultaneous and sequential auctions is of great practical concern, as an auction originator with multiple objects, or multiple units, to auction will need to decide on whether to run one auction or more, and if more, how many.

The auction practice sections address measures that can be taken to mitigate bidder collusion and to enhance competition, provisions for information disclosure, volume adjustments, and other basic principles of auction management that can enhance prospects for participation and a competitive outcome. Information disclosure provisions need to balance the benefits of information pooling and the risks of collusion. While full disclosure of all bids can invite market division and coordinated bidding, limited disclosure can often achieve the benefits of full disclosure without increasing the risk of anticompetitive bidding behavior. Volume adjustments are another new administrative tool that can be used to encourage more competitive bidding. These issues are discussed in more detail below. This part of the book also contains some discussion of auction experiments and simulations. Experiments, on one hand, are an increasingly popular research area in which subjects are recruited, and compensated, for participating in auctions staged in controlled environments. Simulations, on the other hand, rely on computer bidders and allow the computer to calculate the outcomes. Both forms of tests have benefits as well as limitations.

The third part of this Primer discusses experience. Most of the discussion surrounds energy and telecommunications auctions, as there is enormous experience in these sectors, and they have been the proving ground for many new auction designs. These sectors also include some of the largest and most newsworthy auctions. This part will discuss auctions in other sectors, including natural resources, commodities, and general procurement.

2 Game Theory, Auction Design, and Strategy

Summary

Game theory provides analytical techniques needed to analyze auction design and strategy. This is a relatively nontechnical introduction. The chapter focuses on how to apply game-theoretic techniques to the analysis of bid strategy and auction designs.

2.1 Game Theory and Auctions

Auctions have specific rules; these rules govern participation requirements, the form of bids, winner determinations, and payments. The rules of an auction can almost always be translated into a precise mathematical formulation. Such formulations are what is more formally called a mathematical *game*. A mathematical game is a representation of set of strategic interactions, similar to those that occur in ordinary games, into mathematical terminology. Such representations can facilitate analysis of the strategies and outcomes, or *equilibria*.

This does not mean that it is always easy to *solve* an auction, as one would solve a system of equations. It not always easy, or even possible, to find or write a computer program even to compute auction outcomes. However, mathematical descriptions can lead to algorithms that solve part or all of an auction. Often it will be possible to capture the algorithms in computer programs or solutions of systems of equations, or to identify what specifically needs to be solved. Numerical techniques can also be applied to check for consistency of any planned strategy or desired goal of an auction design, using the underlying mathematical structure of an auction.

This chapter begins by describing what constitutes a *game*. Then some basic solution concepts for games are presented. It is explained how an auction can be interpreted as a game, and how the solution concepts from noncooperative game theory can be used to identify equilibria, likely outcomes, of auctions.

2.2 Noncooperative Games

An auction is generally best modeled as a *noncooperative* game. This does not mean that bidders will not cooperate in an auction; it only means that the bidders can have conflicting and/or competing interests. Auctions can be single-stage or multi-stage or multi-round. Single-stage and multi-stage games are represented in two distinct ways.

2.2.1 One-Shot Auctions

Many auctions are one-shot games. For example, in a sealed-bid auction, bidders name prices, and no bidder has any chance to respond. Sometimes sealed-bid auctions allow bids to include not only an offer but also desired quantities and nonprice attributes. Usually the high bidder wins, and the amount paid is the bid amount. But other payment rules are common.

2.2.2 Normal Form Games

A single-stage game is most often described in what is called *normal,* or strategic, *form*. A normal form game is defined by a triple: a set of players $N = (1, 2, \ldots, N\}$, a strategy set of each player, S_j, $j \in N$, and a payoff function for each player, $\pi_j(s_j, s_{-j})$, describing player or bidder j's payoff or profits as a function if its strategy s_j and that of all the other players, s_{-j}:

The prisoner's dilemma is a classic example of a noncooperative game. In it the row player's strategy set is {Top, Bottom} and the column player's strategy set is {Left, Right}. The payoffs are given by the entries in the payoff matrix shown in table 2.1.

There are many types of these matrix games. In the prisoner's dilemma, the players have *dominant strategies*[1] of always playing the

Table 2.1
Prisoner's dilemma

Row/column player	Left	Right
Top	(1, 1)	(0, 3)
Bottom	(3, 0)	(2, 2)

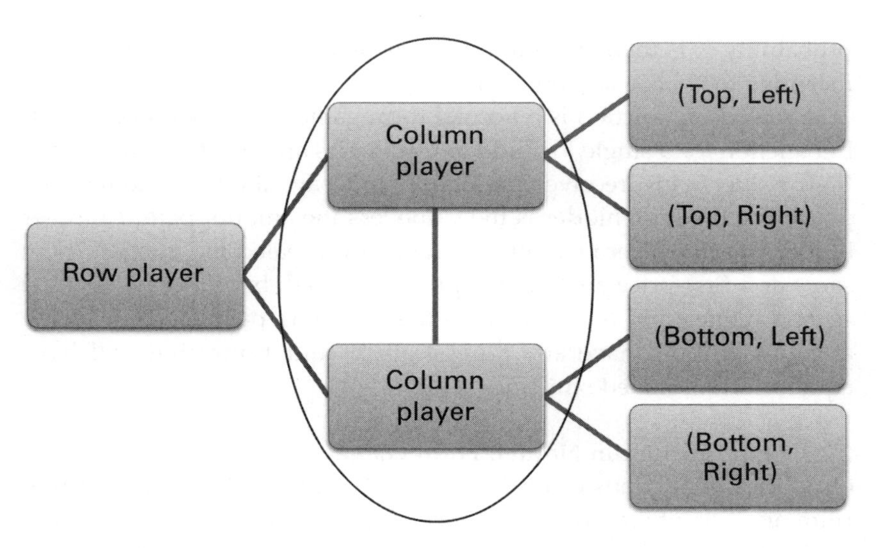

Figure 2.1
Prisoner's dilemma in extensive form

strategy that leads to a joint payoff *minimum*. Another basic example of a normal form game is a *strictly competitive* game, such as the matching pennies game. In the matching pennies game (table 2.2) each of two players can choose either heads or tails. If the choices match, then the first player wins both pennies, and otherwise, the second player does. In this game, if one player does better, the other does worse, and that is why it is called a competitive game. In this game there is no equilibrium in *pure* strategies. If the first player were to choose heads, then the second player would want to choose tails, and vice versa. Similarly, whatever is the choice of the second player, the first player would want to match that choice.

In order to characterize what players would do in this type of setting, one must consider random, or *mixed*, strategies. A mixed strategy for the row player in the prisoner's dilemma is a probability of choosing Top, and similarly a mixed strategy for the column player in the prisoner's dilemma is a probability of choosing Left. In the matching pennies game, if player 1 were to choose heads with probability $\frac{1}{2}$, then player 2 would be indifferent between choosing heads and choosing tails, and similarly, if player 2 were to choose heads with probability $\frac{1}{2}$, then player 1 would be indifferent as to whether it chose heads or tails. This type of randomization, each player choosing heads with probability $\frac{1}{2}$, is then an equilibrium, as neither player would have an incentive to choose any other strategy.

A sealed-bid auction is a normal form game. In a first-price sealed-bid auction for a single lot, bidders' strategies are prices for the lot; the high bidder wins, receives the object, and pays the bid amount. The payoff for the high bidder is the value less the amount paid. All other bidders receive a zero payoff. A second-price sealed-bid auction is the same as a first-price auction except that the high bidder only pays the second highest bid amount. As in the matching pennies game, some auctions will not have any pure strategy equilibrium but will have equilibrium in mixed strategies.

2.2.3 Equilibrium in Normal Form Games

Outcomes of auctions can often be readily predicted through equilibrium analysis of simple normal form models. The most basic form of equilibrium is called a Nash, or noncooperative, equibrium. A *Nash equilibrium* is a set of strategies, one for each player, such that no player can make a unilateral change in its strategy and increase its payoff. A Nash equilibrium is a minimum requirement for each entity to be acting consistently in line with each entity's view of its own interests.

One alternative, and equivalent, definition of a Nash equilibrium, for noncooperative games in general and in a matrix game, such as the prisoner's dilemma in particular, is that each player has an expectation about what its rivals are doing, its choice of strategy is a best reply to those expectations, and every player's expectations are the same as the actual strategies chosen.

In the prisoner's dilemma example, the only Nash equilibrium is (Top, Left). This is clearly suboptimal for the players. Indeed this outcome is a *dominant strategy equilibrium*: each player's choice of strategy is the same, no matter what the other chooses.

A multi-unit auction can have a prisoner's dilemma component. Suppose, for example, that there are two blocks available, and each of the two high-value bidders has declining marginal values, say H for the first block won and $M < H$ for the second block won. Also suppose that other bidders have values of no more than $L < M$. Then it will be an equilibrium for each firm to keep bidding in an ascending price auction up to M for a second block, even when their both stopping at L would result in higher surplus and the same allocation for the two high bidders.

In the matching pennies game there is also a unique equilibrium: for each player to choose heads with probability $\frac{1}{2}$. This is a mixed strategy

Table 2.2
Matching pennies

Col\row player	Heads	Tails
Heads	(+1, –1)	(–1, +1)
Tails	(–1, +1)	(+1, –1)

equilibrium. It is not a dominant strategies equilibrium, as the optimal choice for one player depends on what other players choose:

2.2.4 Mixed Strategy Equilibrium in Spectrum Auctions: Some Examples

This section provides a few sample characterizations of Nash equilibria in some actual spectrum auctions. These cases are examples of actual auctions that only have mixed strategy equilibrium.

The first example is an auction of 2G spectrum that the Dutch government was planning to run in 1998. The government changed the rules after hearing bidder concerns. In that auction there were 16 very similar blocks of spectrum for sale, and bidders were to be asked to submit a final round of separate sealed bids for each of those blocks. Bidders would need to win between 4 and 6 blocks in most cases to derive any value. Bidders submitting bids on only 6 blocks could end up winning fewer than the necessary 4, suffering from an exposure problem. And bidders who wanted to maximize chances of winning at least 4 blocks, could submit bids on all 16 blocks, and end up winning many unneeded blocks. A simplified version of this game was analyzed extensively by Szentes and Rosenthal (2003), in which two bidders are competing for three blocks and each has to win at least two to derive any value.[2] They were able to completely characterize the equilibrium probability distribution for the mixed strategy of this auction.

Another example of auctions which often have no pure strategy equilibria comprises the *supply function* auctions commonly used in energy procurement. These auctions require bidders to submit a separate price for each block (a step function bid).[3] The bidders' offers are aggregated to form a supply schedule, and the lowest offers that meet demand are the ones that are selected. In a *paid as bid* auction, bidders will receive their bid amounts for the portion of their supply that is selected by the auctioneer.[4] The prices each bidder receives can be uniform, or paid as bid. In a uniform price auction all winning bidders receive the market-clearing price.

2.2.5 Equilibrium in Multi-Attribute Auctions

At times, the auction originator's objectives will include nonprice factors. This is especially true when the auction is being run by a government agency that has a mandate to promote the public interest. Many spectrum auctions include coverage or buildout provisions. And at times, bidders receive a score based on price and nonprice submissions. For example, a recent French auction for 4G spectrum licenses weighted each bidder's monetary offer by factors that depended on whether the bidder would cover rural areas and provide wholesale access to resellers.[5] The weights for the nonprice attributes, which were 0 or 1 decisions for the bidders, were high. This resulted in all winning bidders including these attributes in their offers.

Because bidders will often try to maximize net returns or profits, when both price and nonprice attributes contribute to the winning score, an equilibrium bid strategy will often be a *corner* solution; in other words, bidders will submit the maximum or minimum allowed values for the nonprice component of the bid, according as the bidder weighs the attribute more or less than the auctioneer.

One extreme example of this property of multi-attribute auctions was a California procurement of *qualifying facilities* (QF) in a sealed-bid auction. The auction was designed to ensure supply of renewable energy sources. Numerous factors were included in determining the score of each bidder.[6] The auctioneer was seeking to procure electricity from renewable resources for both peak and off-peak periods, and to promote the construction of renewable generation capacity. Bidders were required to submit separate prices for different components of the generation service: one was a price required for ensuring capacity availability, and others included prices for energy supplied in different periods. Wind farms were among the most significant sources of renewable energy. However, wind turbines only operate when there is wind. Therefore the maximum capacity of a wind farm (or of other renewable energy source such as a hydroelectric facility) may not represent average availability. In this case the auctioneer wanted to provide different weights on potential availability and actual production. The weight given to capacity in the score was relatively high; the result was that the winning bidders ended up receiving high-capacity prices and *negative* energy prices. In other words, these wind farm developers received less money the more energy their wind farms supplied.

If the auctioneer can measure the monetary value of the nonprice attributes, then using those weights will provide bidders the appropriate incentives to submit optimal combinations of price and nonprice attributes. Simple Nash equilibrium analysis of bidder incentives when bids include price and nonprice attributes will usually suffice to anticipate the type of pitfalls that were experienced in the California QF auction.

2.3 Multi-Stage and Sequential Auctions

In many instances an auction will have multiple stages, so that bidders can respond to rivals' bids. There may even be a sequence of auctions for substitutes, for complements, or over time. Equilibrium criteria for one-shot games cannot allow for interactions and information revealed between stages or auctions. The *extensive form* representation of a game can be used to capture these effects.

An extensive form representation of a game includes three components: a game tree, decision nodes, and information sets. The tree describes the possible sequences of decisions and events, which can include random events. The decision nodes are points at which a player can take an action. And the information sets represent the information available to a player at each of the decision nodes. The graph of the prisoner's dilemma in figure 2.1 shows how a normal form game can be represented in extensive form.[7]

Some auctions, such as the Federal Communications Commission (FCC) spectrum auctions, allow bidders to gradually improve their offers over the course of the auction. These multi-stage (or multi-round) auctions can be represented in extensive form.

Other times, for example, in energy markets, the same bidders compete repeatedly over time. This type of auction can be characterized as a dynamic game, and can be represented in a variation of the standard extensive form, which is illustrated above.

2.3.1 Subgame Perfect Equilibrium

Multi-stage games differ from one-shot games in that the players interact repeatedly in the former and only once in the latter. After each round of interaction, the players can revise their plans. The notion of a subgame perfect equilibrium (SPE) is intended to ensure that equilibrium strategies remain so through the entire course of play. More specifically, a *subgame perfect equilibrium* in an extensive form game is a Nash equilibrium with the additional property that the equilibrium strategies remain in equilibrium when restricted to any proper subgame, that is, any intermediate node of the game in which players know the prior history. What this means is that after each stage, round, or period, when a player has to make a decision knowing previous decisions of other bidders, the bidder will not want to revise its strategy at that point.

As an example, consider an auction for three blocks, 1, 2, and 3; suppose that there are two strong bidders, A and B. The bids for each block are collected after the winner and price for the previous block is determined. (This is equivalent to the situation in which bids can be collected all at once, opened one at a time, and each bid can be contingent on whether or not a bidder won one of the previous blocks.)

Suppose that the price for each block is the best losing bid for that block. Say both A and B prefer block 1, and A is the only bidder who has a high value for block 3. Let blocks 1 and 2 be very close substitutes, and let bidder A be the stronger bidder. Then the following strategies are a Nash equilibrium, but not subgame perfect: bidder A bids low for block 1, high for block 2, and high for block 3, contingent on winning one of blocks 1 and 2; bidder B bids high for block 1, low for block 2, high for block 3 if it wins block 2, and low for block 3 if it wins only block 1.

B's high bid for block 3 in the event that it loses block 1 is a noncredible threat to deter A from competing for block 1. What is a SPE is for B to bid low for block 1 and high for block 2, and vice versa for A. If B were to lose block 1, it would have no incentive to carry out its implicit threat.

Further stability requirements can, at times, further refine the notion of a SPE.[8]

2.3.2 Signaling Games

In many multi-stage auctions, bidders can signal interests or intent.[9] For instance, in the United States the FCC has conducted several dozen multi-round spectrum auctions. In these auctions early round bids are often well below final prices, and these bids do not represent any commitment by bidders, as the bids are low and certain to be topped in later rounds. In such auctions, concerns have been raised that bidders can signal so as to achieve coordinated outcomes.[10] The question addressed here is to what extent one can model such behavior as an equilibrium outcome. Contrary to what may be popular opinion and some literature, collusive signaling need not have any effect on the outcome. This is not to suggest that signaling in auctions has not occurred and/or has been ineffective. Rather, signaling is as likely to be ignored, or even counterproductive, as it is to be effective.

To see this, consider a simple situation in which there are two bidders, A and B, and two blocks, 1 and 2. Suppose that each wants one block, and both prefer block 1, but both prefer block 2 to having to compete for 1. So there can be two equilibrium outcomes: 1 goes to A and 2 to B, and the reverse. Now suppose that before bidding each bidder can announce which block it will bid for. Presumably, if A announces 1 and B announces 2, then that should be the outcome. However, neither bidder may trust the other, and either can misinterpret the other. Absent an ability to form an ex ante binding agreement, these announcements are cheap talk, and can be ineffective or worse.[11]

Indeed one case that Cramton and Schwartz (2000) cite about how bidders can signal each other in spectrum auctions shows how such efforts can backfire. In that auction, bidders sent signals—in the form of jump bids and overbidding—that two bidders were able to convey, observe, and (apparently) understand during the auction. However, one of the bidders, the *receiving party*, did not really want to coordinate in the way that the *sending party* signaled. During the auction, the sending party could only see that the receiving party was apparently acting cooperatively, and not that the receiving party had contacted the regulator. After the auction, the sending party was subject to prosecution and fined for attempting to collude during the auction.

2.4 Repeated Games

At times, the same bidders compete repeatedly for more or less the same lots. An example of this is the daily auctions conducted on most power exchanges.[12] The simplest form of repeated interaction occurs when the same game is repeated over and over. The notion of a SPE can be extended to these repeated games. As infinitely repeated auctions have no final auction, rivals' bids in subsequent auctions can depend on the outcome of a given auction, and can drive bidding in virtually any direction. This means, when the game is repeated indefinitely, there are infinitely many possible SPE outcomes.

In contrast, for many repeated games, including all those with a unique equilibrium in the one-shot stage game, there will be only one equilibrium in the finitely repeated game, no matter how long the interaction lasts, as long as there is a fixed end date and players are rational.

What follows is a brief description of the types of outcomes that can arise in repeated interactions. The discussion assumes that exactly the same game is repeated. However, most of the analysis applies to more general dynamic games.

2.4.1 Finitely Repeated Games

Consider the case where the prisoner's dilemma, as in table 2.1, is repeated a fixed and finite number of times. Suppose that at each stage, or iteration, the row player chooses Top or Bottom, and the column player chooses Left or Right. These choices are made simultaneously. After the choices are made, the players are each informed about the other's choice, payoffs for the stage are accrued, and then the process repeats.

One could consider the possibility that each player will, in early stages, play cooperatively, and do so at least until near the end, as long as the other player does so. This is not a SPE, as each player can infer that the other will play noncooperatively (i.e., Top or Left) in the last period. Therefore each player has an incentive to play Top or Left in the next to last period too.

The same logic will then apply to the third to last stage, as each player knows the other will choose to play noncooperatively in the second to last stage. By backward induction, there will be only one SPE in the entire game, which is for the row player to always choose Top and the column player Left.

If the game has a probability p of running an additional period, so that the end is uncertain, then other outcomes are possible. More specifically, if either player defects from a cooperative strategy, it will earn 3 for one period, and 1 for all subsequent periods—for however many periods the game lasts. The present value of this strategy is $3+\delta p/(1-\delta p)$, where δ is the discount factor. In contrast, continued cooperation results in an expected payoff of $2+2\delta p/(1-\delta p)$. And this quantity is larger than the first whenever $2\delta p > 1$, that is, when discounting is not too fast and the probability of the game ending fast not too high.

So, for example, in repeated interaction of traders in energy markets, there is a possibility of the traders all refraining from aggressive bidding, even where there is uncertainty. When there are multiple Nash equilibria in the stage game, cooperative outcomes can still be stable with finite interaction, provided that there are bad and good equilibria and that each player wants to avoid the bad equilibrium in the last stage.[13]

2.4.2 Infinitely Repeated Games

An infinitely repeated game is one in which the same players interact in the same strategic situation indefinitely. The main, and big, difference between finitely repeated games and infinitely repeated games is that there is no last period. This means that unlike a finitely repeated game, the backward unraveling from cooperative Nash equilibria does not occur. Therefore cooperative outcomes are possible. Indeed, in infinitely repeated games, it can be shown that any payoff combination in the stage game that dominates the lowest payoff that each player can guarantee itself (its minmax payoff) can be supported using *grim trigger strategies* as a SPE, as long as the discounting is not too fast. A grim trigger strategy is one in which each player adheres to the cooperative play as long as no other player has deviated in any previous round. If there was a deviation by a player, then everyone reverts to that strategy that gives the defector its minmax payoff thereafter. When discounting is slow enough, the one-shot benefit from a deviation is outweighed by the benefits of continued cooperative play.[14]

This means that the outcome of repeated auctions can be indeterminate. As is discussed in more detail below, information that limits the ability of bidders to coordinate actions can induce bidding to be more competitive. This indeterminacy result, called a *folk theorem* for repeated games, can be extended to situations in which players cannot directly observe rivals' actions. When actions are not directly observable, payoffs are still observable, and so a player might become increasingly certain over time that a rival or rivals are deviating from cooperative play. Nevertheless, cooperative play can be a SPE when each player uses a stochastic decision rule: continue to cooperate as long as the price, or payoff, has averaged at least a threshold level, and if not, then revert to the Nash equilibrium (*punishment*) for a fixed and finite number of periods.[15]

The reason to use a stochastic decision rule is that demand or cost shocks can result in low prices or profits even when there are no defections. So the rule to use for determining whether to continue to cooperate has to leave some buffer for uncertainty. The reason for a finite punishment duration is that there is always a positive probability of triggering the punishment, and for that reason the equilibrium should leave room for reversion to cooperative play.

2.4.3 Overlapping Generations of Players

The assumption that play goes on forever is consistent with many institutions, but not with the horizons of the individuals making decisions. However, there are generally older and younger players involved. As long as it is commonly accepted that older players will tend to act more opportunistically and that the younger players will get their turn, then cooperative play is still a SPE. More specifically, a folk theorem result still applies when (1) there is an infinitely repeated game in which the identities of the players change over time, (2) each player only considers its payoffs during its term, so that there are no bequest motives, (3) there is no period in which every player leaves and is replaced by a younger player, and (4) terms are long, and discounting is slow enough. Should a younger player not tolerate an older player getting a larger share, then the entire play would revert to competitive play for the rest of that player's term. This is enough to support cooperative behavior.

2.5 Summary

This chapter has explained how auctions can be modeled as noncooperative games. Single-round or single-stage auctions can be modeled as normal form games, and the Nash equilibrium concept is useful in analyzing such auctions. At times the Nash equilibrium solution will not be able to specify certain outcomes but rather random ones (mixed strategy equilibria). Such equilibria can be useful in explaining what might otherwise appear to be inconsistent behavior in some auctions.

Extensive form games can be used to represent sequential auctions. Subgame perfect equilibrium will often characterize the plausible outcomes, and ones that rule out noncredible implied threats. However, there can be room for signaling in extensive form games to further refine the equilibrium set. Such signaling will not always be effective, even when there are no costs of conveying signals that can be understood by rival bidders. Costly signaling can nevertheless be used to sort out different types of outcomes, such as those that might be achieved if everyone knew certain bidders were strong, from those that would be achieved if those same bidders were known to be weak.

SPE of repeated games can lead to a wide range of outcomes. Auction rules that limit information available to bidders can make bidding more competitive. However, absent some model of behavior, it can be difficult to predict what type of outcome is likely in repeated auctions.

3 Revenue Equivalence

Summary

This chapter explores the relations of outcomes in different auction types. It shows that a first-price sealed-bid auction is strategically equivalent to a Dutch auction. Also conditions are provided under which an English auction is strategically equivalent to a second-price sealed-bid auction. This chapter then provides an explanation of the derivation of perhaps the single most significant result in auction theory, the revenue equivalence theorem (RET). This result states that the expected revenue and bidder payoffs from two auctions will be the same when the final allocations are the same.

3.1 The Four Basic Auction Types for Single-Object Auctions

A single auction can include one or multiple objects, and can also include multiple units of each object available. Multi-object auctions, and even multi-unit auctions for a single product, confront bidders with the problem of anticipating prices in other auctions or for other objects when bidding on any single unit or object. This is explicitly the case in a sequence of auctions for substitutes and/or complements. But it is also the case when there are parallel independent auctions. This chapter focuses on single-object auctions, where these interdependences are not an issue. Much insight can be derived from an analysis of these types of auctions. A great many different auction formats have been used even for single-object auctions. By far the most common are the four main traditional types of auctions:

1. English, ascending price, open outcry
2. Dutch, descending price

3. First-price sealed-bid
4. Second-price sealed-bid

The English and first-price sealed-bid are by far the two most common auction formats. These auctions are also primarily used for the sale or purchase of a single object at a time. Each of these auctions can be used iteratively to sell multiple objects. This is typically the case at both traditional on-site auctions and in many online auctions. However, as will be seen in subsequent chapters, these four traditional auctions are not well suited for the sale of substitutes or complements, which involve strategic cross-auction interdependences for bidders.

The English auction has several, closely related variants. One version has the auctioneer announcing an ascending sequence of prices, and bidders agreeing to keep in the auction or dropping out.[1] A strategy for a bidder in this form of the English auction is a rule to determine when to stop bidding (i.e., drop out). In a truly open auction, bidders will see the number of other bidders still remaining, and possibly their identities, when making a decision about whether to drop out. This can be material information and, as is explained in further detail below, can materially affect the outcome.

More commonly, auctioneers tend to start the price very high, then quickly backtrack and lower the starting price. In addition, bidders do not have to agree to each announced price (except in the Japanese auction variant), and at times bidders can call out their own price suggestions with larger or smaller increments than is suggested by the auctioneer. Online auctions typically also have the feature of the English auction that bidders are asked to respond to an increasing sequence of prices. However, online auctions typically have fixed deadlines for submitting bids.[2] These deadlines mean online auctions are strategically much different than the traditional English auction. Nevertheless, when the online auction has an extension rule that prolongs the auction by a fixed amount of time after each new bid, the auction is essentially an English auction.

The descending price Dutch auction has the auctioneer start with a high price, and the auction ends when the first bidder enters the auction.

The first- and second-price sealed-bid auctions are auctions in which each bidder submits a sealed bid for the object for sale. In both, the high bidder wins. In the first-price auction, the winner pays the bid amount. In contrast, in the second-price auction, the winner pays the highest losing bid amount.

One of the main insights of the early auction literature[3] is that the English and the second-price sealed-bid auctions are, in many, but not all, cases, strategically equivalent. On the other hand the Dutch auction and the first-price sealed-bid auction are always strategically equivalent. What is meant by strategically equivalent is that the strategies bidders would want to use are fundamentally identical.

As is explained below, this strategic equivalence of the English auction and the first-price auction will hold when information revealed during the course of bidding in an English auction will not affect a bidder's valuations or strategies. However, when information a bidder gets during bidding rounds can affect valuations or expectations, then bidding behavior is affected, and the strategic equivalence need not hold.

3.2 Auction Strategy in the Four Basic Auction Types

Each of these four auctions can be described as a normal form game. Sealed-bid auctions are very straightforward to describe in this fashion. In both first- and second-price auctions, the bidders (the set of players) have as a strategy the choice of a monetary amount to offer for a single available object. The winner determination in both is the same—the high or best bidder (assuming a forward auction). The only difference is in the payments: in the first-price auction, the winner pays its bid amount. In contrast, in the second-price auction, the winner pays the second highest bid amount. Losers pay nothing.

The Dutch auction is also very straightforward to describe in normal form. In a Dutch auction, the bidders (players) will see a decreasing sequence of prices. The prices may decrease more or less continuously, or in fairly large steps. A strategy is the price at which to first agree to purchase the item. The winning bidder is the one that enters first and pays its bid amount. Losers pay nothing.

In an English auction the strategy can be more complicated. Bidders face an increasing sequence of prices. A strategy is a decision rule determining when to drop out. The decision rule can depend on the size of the increment and on the number of other bidders remaining in the auction at each price. This decision can also depend on how fast each rival bidder is increasing its offer. In simple auctions, when prices increase continuously or nearly so, the bidders will base the decision to stop bidding on their own values, and not on the size of the increment. If the outcome of other bidders is immaterial or unobservable, then the strategy for when to stop bidding is simply a limit price. In more complex auctions in which rivals' information or identities are material, the bidding strategy can depend on who is still bidding in the auction. The winner is the last bidder in, and at the last agreed-on price.

The Japanese auction variant of the English auction requires all bidders to remain in at each posted price as long as they are willing to purchase the item. Once a bidder drops out, it is no longer able to reenter the bidding. As in the standard English auction, the activity of other bidders can materially affect any given bidder. But, if not, the strategy in the Japanese auction is simply a stopping price.

3.3 Strategic Equivalence

This section explains the strategic equivalence of first-price auctions and Dutch auctions, and of second-price auctions and English auctions.

First consider the English auction. As noted above, in the most usual English auction, an auctioneer calls out an escalating sequence of prices, and at least one bidder that is not the standing high bidder must meet the most recently announced price within a fixed amount of time to keep the auction from ending. In this type of auction, the optimal strategy for a bidder with a value v is to keep bidding as long as the price is no higher than v, and then drop out.[4] This assumes that bid increments are small; if not, then a bidder may have an incentive to skip a turn responding to a new bid. This also assumes that a bidder does not revise its valuation during the auction based on what it sees from others' behavior. This is not always the case.

For example, in a common-value auction, each bidder has a signal of the true value $s_j = v + \epsilon_j$, where v is the true value and ϵ_j is an independently, identically distributed error term. This type of situation is common in oil lease bidding.[5] In this case information about others' bids provides some information about a bidder's forecast error. In the case of oil lease bids, it was observed that although bidders may have, on average, unbiased forecasts, they will only win when their forecasts are too optimistic. Seeing other bidders staying in or dropping out provides information about how optimistic is a given bidder's forecast; this information can be used to revise expectations. This type of revision is not needed in an auction in which bidders have independent private values, that is, each bidder's value is uncorrelated with its rivals'.

In the second-price sealed-bid auction, it is straightforward to see that a bidder's dominant strategy is to bid its true value.[6] No bidder has any incentive to bid more than its value. Suppose that a bidder offers $b > v$, and the highest offer by a rival is b'. There are three cases to consider:

1. $b' \leq v < b$. Then overbidding does not matter. This bidder wins and pays b'.
2. $v < b' \leq b$. Then this bidder wins, and pays $b' > v$. In this case, overbidding reduces this bidder's payoff.
3. $b' > b$. Again, overbidding does not matter, as this bidder loses.

A similar argument can be used to show that a bidder will never want to bid any $b < v$. If it did, there is a chance it would lose, when it should not. Still, if bidding $b' > b$ does not affect whether the bidder wins, then this higher bid would impose no additional costs on the bidder. Thus underbidding cannot be an equilibrium. This shows that in a second-price sealed-bid auction, it is always optimal to bid one's value; in other words, this is a dominant strategy.

What this means is that the point at which a bidder stops bidding is the same in an English auction as the amount a bidder would offer in a second-price sealed-bid auction. Further the bidder pays the second highest bid amount. Thus the auctions are strategically equivalent, and achieve the same prices and payoffs to the bidders. The auctions also give the seller the same price, and the same buyer pays it. Thus these two auctions are also *outcome* equivalent, as well as *payoff* equivalent, at least in an independent private values setting. Note that if bidders cannot observe rivals' bids, and price increases are continuous, the fact that the auction continues can convey information. When bidders are never told who else is bidding, or how many other bidders remain active, and the stopping rule is for the auctioneer to continue increasing price until a fixed upper bound higher than market clearing, then the English and second-price auctions will always be equivalent—strategically and with respect to payoff and outcome.

In a first-price sealed-bid auction a bidder will want to bid less than its value. If a bidder offers its full value, then it will earn zero surplus even when it wins. The calculation of how much a bidder should offer involves a trade-off—a lower offer provides a higher surplus if the bidder wins, but a lower probability of winning. The bidder will want to choose its bid amount b so as to maximize

$$P(b)(v - b), \tag{3.1}$$

where v is the bidder's value and $P(b)$ is the probability that it wins.[7]

In a Dutch auction a bidder will not want to bid until the price falls at least to its value v. Once the price falls to v, the bidder will want to choose a price to make a first offer b so as to maximize $(v-b)P(b)$. This is the same expression as (3.1). Thus a bidder will want to choose the same bid amount in both first-price sealed-bid auctions and Dutch auctions. The Dutch and the first-price sealed-bid auctions are strategically equivalent: bidders will have incentives to bid the same amounts.

In summary:

Theorem 1 (Riley and Samuelson 1981; Vickrey 1961) *In (reduced) normal form, Dutch auction and first-price sealed bid auction are strategically equivalent.*

A strategy in the first-price sealed-bid auction is just the price the bidder offers. In a Dutch auction the strategy is the price at which to start bidding. Assuming that price falls continuously in the Dutch auction, the expression (3.1) establishes that these are the same decisions.

3.4 Revenue Equivalence of English and Dutch Auctions

One question often raised is: What difference does auction design make? For instance, there has been much, and sometimes heated, debate about the relative merits of uniform price versus pay-as-bid auctions in the energy sector.[8] This section explains one of the more remarkable theoretical results of economics—the *revenue equivalence theorem* (RET). The result is that the expected payments and outcome of the English and Dutch, and therefore those of the first-price and second-price auctions, are the same—at least when bidders have independent private values.

Recall that in an English auction, or a second-price sealed-bid auction, all bidders bid truthfully; in equilibrium, the high-value bidder wins and pays (approximately) the valuation of the bidder with the second highest valuation. In the Dutch auction, or the first-price sealed-bid auction, the high bidder will still usually win,[9] but the amount of the winning bid, which is the transaction price, will generally be less than the value. The question is by how much the winning bid amount will fall short of the winning bidder's value. The remarkable answer is that this gap is exactly equal to the expected value of the second highest value of the bidders, or the equilibrium price in an English auction—at least in the case where bidder values are independently distributed.

The intuition for this result is fairly straightforward, although the derivation is somewhat more complicated than for strategic equivalence of the bid strategies in the Dutch and first-price, sealed-bid auctions. Consider the first-price sealed-bid auction. The high bidder won't know ex ante whether it is the high bidder. However, this high bidder should only be concerned with the case where it is the high bidder, as in other cases its bid won't matter.

So the high bidder will want to bid just a bit more than the second highest bid amount. This amount is no more than the second highest value, as the second highest bid can never be above value in equilibrium in a first-price sealed-bid auction. Of course, the second highest bidder will offer a bit less than its value too. So the high bidder will want to weigh the benefit of bidding a bit more than the expected amount of the second highest bidder's value. This high bidder would gain a bit of surplus by bidding a bit less, but would lose with a higher probability. When the high bidder's offer exactly equals the expected amount of the second highest bidder's value, then the benefits of increasing its bid a small amount will exactly equal the costs.

This RET theorem can be established in two different ways. The more direct approach is to compute optimal bid functions, and then to show that the optimal bid in equilibrium means that the high bidder will bid the expected amount of the second highest bidder's value. The next section develops an approach to calculating an optimal bid function but does not provide a complete proof the RET.

The less direct, and much more general, approach is to rely on what is called the *revelation principle*, to restrict attention to *direct* revelation mechanisms, that is, auctions that essentially ask bidders to report their type. In direct revelation mechanisms, the bidders' reports determine the winner and the payment. Auctions are a specific form of direct revelation mechanism—bidders are asked to report values, and payments are based on those reports. Of course, in a first-price sealed-bid auction a bidder won't wish to offer its full value, but its offer can be used to impute its value. In that sense a first-price sealed-bid auction is a direct revelation mechanism.

The RET states that any two auctions that always have the same winners, and give the lowest value bidder the same (zero) expected surplus, always result in the same auction revenues. The intuition for this result is as follows.

Consider a bidder with value v, who submits a bid $b = \beta(v)$, where $\beta(v)$ is a bid function. Now consider the incentives of this bidder to report that it is a slightly stronger or weaker bidder by reporting $b + \Delta = \beta(v + \delta)$. It has to be the case, at equilibrium, that this bidder will not gain by reporting a bit more or less. So the extra payoff is offset by the extra probability of losing by altering a bid. This would have to be true for any two auctions that always result in the same winners. This means that the shape of the total payment schedule, $P(v)$, would have to be same for *any* two auctions that always result in the same winners. And if the lowest value bidders always get a zero payoff, then the two auctions would always generate the same revenues. Klemperer (2004) provides a more general statement.

Theorem 2 (Klemperer 2004, p. 43) *Assume that each of N risk-neutral potential buyers has a privately known value independently drawn from a common distribution $F(v)$, which is strictly increasing and atomistic on $[\underline{v}, \overline{v}]$. Also assume that no buyer wants more than one of the k available and indivisible objects. Then any auction mechanism in which (i) the objects always go to the k buyers with the highest values and (ii) any bidder with value at $\underline{v}$ expects zero surplus yields the same expected auction revenue and results in a buyer with value v making the same expected payment.*

Sketch of Proof

Suppose that values are iid draws on $[\underline{v}, \overline{v}]$. Let $F(v)$ be the distribution, with $F(\underline{v}) = 0$ and $F(\overline{v}) = 1$. Let $S_j(v)$ be the expected surplus of bidder j with value v. Let $P_j(v)$ be the probability that bidder j wins.

So $S_j(v) = vPj(v) - E$[payment by type v of player j]. This implies that if bidder j reports any $v' \neq v$, then $S_j(v') = P_j(v')v$ – [expected payment by a bidder j of type v'].
So $S_j(v) \geq S_j(v') + (v - v')\, P_j(v')$. This implies that $S_j(v + dv) \geq S_j(v) + (-dv)\, P_j(v + dv)$. Also, $S_j(v + dv) \geq S_j(v) + dvP_j(v)$.
The two inequalities above imply that $P_j(v + dv) \geq [S_j(v + dv) - S_j(v)]/dv \geq P_j(v)$. Letting $dv \rightarrow 0$ leads to $dS_j/dv = P_j(v)$. Note that if $S(v) = vP(v) - X(v)$, where $X(v)$ is the payment that the bidder must make, the condition becomes

$$\frac{d[v_j P(v) - x(v)]}{dv} = P_j(v).$$

Integrating up to v yields

$$S_j(v) = S_j(\underline{v}) + \int_{\underline{v}}^{v} P_j(x)dx.$$

Now consider any two mechanisms that have the same $S_j(v)$ and the same $P_j(v)$ for all v and for all players j. So j's expected payoff is the same for both mechanisms. This means that the auctioneer gets the same expected revenues. ■

A direct application of the result above is the revenue equivalence of first-price, second-price, English, and Dutch auctions when bidder valuations are independent draws from the same distribution. This result is also key to characterizing an optimal auction, as is explained in more detail in the next chapter. Absent any reserve price, the second-price or English auction yields an expected revenue of the value of the second highest valuation among the bidders. So, at least in the case of independent private values, a second-price auction with a positive reserve price will be optimal.

The RET, as stated in theorem 2, makes a few significant assumptions. One is that each bidder is only interested in at most one unit. Should a bidder want more than one unit, strategic withholding, that is, bidding low on one unit hoping to affect the clearing price can be a factor. This possibility is discussed in more detail in chapter 8. Another key assumption is that buyers have independent private values. As is explained in chapter 5, RET no longer holds in a *common-value* auction, that is, an auction in which values of different buyers are correlated.

3.5 Summary

This chapter has considered single-object auctions. The main insight, that the auction rules for determining the winner and the price paid can affect bidder incentives, was used to characterize some key properties of auctions. First, there is a strategic equivalence between Dutch and first-price sealed-bid auctions. Second, under some assumptions, there is also a strategic equivalence between standard English auctions and second-price sealed bids. The more striking result is that the auction revenues can be the same for two auctions that appear quite

Table 3.1
FCC auction 11

Block	Market name	Number of bids	High bidder	High bid ($k)	Net bid ($k)	Price ($/bu)	Average price D/E—price of F	Percent difference
E	New York,	0	AT&TWire	58,800	58,800	0.33		
F	New York,	0	NorthCoast	100,320	75,240	0.56		
D	New York,	0	OPCSE	50,700	50,700	0.28	($20,490,001)	-40%
F	Los Angele	0	AerForce	5,965	4,474	0.04		
D	Los Angele	0	AT&TWire	37,510	37,510	0.26		
E	Los Angele	0	Rivgam	31,910	31,910	0.22	$28,745,000	90%
F	Chicago, IL	0	NextWave	30,753	23,065	0.38		
D	Chicago, IL	0	SprintCom	59,976	59,976	0.73		
E	Chicago, IL	0	SprintCom	62,741	62,741	0.77	$30,605,499	49%
D	San Franci	0	AT&TWire	13,655	13,655	0.21		
F	San Franci	0	NextWave	5,779	4,334	0.09		
E	San Franci	0	Western	10,737	10,737	0.17	$7,861,753	73%
D	Philadelphi	0	Comcast	12,169	12,169	0.21		

Table 3.1
(continued)

Block	Market name	Number of bids	High bidder	High bid ($k)	Net bid ($k)	Price ($/bu)	Average price D/E—price of F	Percent difference
F	Philadelphi	0	NextWave	29,407	22,055	0.50		
E	Philadelphi	0	Rivgam	12,761	12,761	0.22	($9,590,250)	-75%
D	Detroit, MI	0	NextWave	3,815	3,815	0.08		
E	Detroit, MI	0	OPCSE	3,856	3,856	0.08		
F	Detroit, MI	0	OPCSE	8,500	6,375	0.18	($2,363,081)	-28%
D	Dallas, TX	0	AT&TWire	25,895	25,895	0.60		
E	Dallas, TX	0	AT&TWire	27,060	27,060	0.62		
F	Dallas, TX	0	NextWave	21,340	16,005	0.49	$1,695,000	8%
F	Boston, MA	0	NorthCoast	8,909	6,682	0.22		
D	Boston, MA	0	OPCSE	6,515	6,515	0.16		
E	Boston, MA	0	OPCSE	7,515	7,515	0.18	($1,893,996)	-25%
F	Washington	0	AerForce	11,780	8,835	0.29		
E	Washington	0	OPCSE	6,071	6,071	0.15		
D	Washington	0	Rivgam	6,820	6,820	0.17	($5,334,493)	-78%
E	Houston, T	0	AT&TWire	9,835	9,835	0.24		
D	Houston, T	0	SprintCom	13,259	13,259	0.33		
F	Houston, T	0	Telecorp	10,150	7,613	0.25	($1,869,539)	-18%

different, such as the English auction, the second-price auction, the Dutch auction, and a standard first-price sealed-bid tender.

In practice, the RET suggests that discriminatory-price auctions, in which each winner pay its own bid amount rather than same price others pay for identical lots, will not always produce better revenues or results than uniform-price auctions. As noted above, this means that decisions about the payment rule that occur in some sectors, such as energy, should focus on whether revenue equivalence holds and on bidder incentives, and not on the fact that bidders pay less than bid amounts in a uniform-price auction. Also spectrum auctions that provide some bidders with discounts can create strong incentives for bidders to find ways to qualify, with the net effect of bidding away such discounts.

This has been the experience in the US FCC spectrum auctions, where bidding credits have been competed away. The experience in FCC auction 11 illustrates this quite clearly (table 3.1). That auction included three identical licenses in each of 493 geographic regions, called basic trading areas.

One of the three licenses, the F block, was reserved for "designated entities" (small bidders), and the other two blocks were available to all bidders. The small bidders also received bidding credit; that is, they paid a discounted price. The net effect was that the small bidders paid virtually the same prices as the larger ones.[10]

4 Optimal and VCG Auctions

Summary

This chapter provides a description of how the optimal bid strategy is calculated for a bidder. This methodology is then applied to characterize an optimal auction design. The chapter concludes with a characterization of all auctions in which bidders would have incentives to accurately report true valuations, that is, Vickrey–Clark–Groves auctions.

4.1 Optimal Auctions

“Optimal” in this chapter is used to mean an auction design and configuration that maximizes the expected revenues from a forward auction, or minimizes the expected costs from a reverse auction. It turns out that often many seemingly quite different auctions will achieve the same expected outcome. This should be apparent from the revenue equivalence theorem, which states that the English, Dutch, first-price, and second-price auctions all result in the same outcome in the case where all bidders have independent private values. The reservation price is a configuration parameter of all four main auction formats discussed in the previous chapter, and it can affect the outcome: the level of the reservation price can determine whether or not an object is sold. Nevertheless, a public and secret reservation price, under the assumptions of independent private values, should result in the same expected revenues, that all bidders will participate and higher value bidders always offer more than lower value bidders. In practice, bidders face participation costs, and so a positive reservation price can reduce participation and influence the outcome and auction revenues. This is also not to suggest that any two auctions will always result in the same expected revenues, even assuming the same winners and the same reservation prices. As will be explained in chapter 5, when bidders' values are affiliated, the English auction generates higher revenues than does either a first-price or a second-price sealed-bid auction.

4.2 Calculating Optimal Bid Functions

This section shows how to directly calculate optimal bid functions for a bidder seeking to maximize its expected surplus in a first-price auction. This calculation can take two approaches. The first is to rely on RET, and estimate the amount of the second highest bidder's value. This approach works for Dutch auctions and first-price auctions when bidders have independent private values. The second, more general, approach is to directly optimize the expected bidder surplus.

The following describes how optimal bid strategies can be calculated for the case where there are N symmetric bidders competing in a first-price auction. Each bidder j's valuation, v_j, is, ex ante, random. The density function of the realized valuations is the same for all bidders, is denoted $f(v)$, and the corresponding distribution is $F(v)$. A strategy is a function $\beta(v)$ mapping a bidder value into bids. Then, in this symmetric case, each bidder of type v will choose a bid b to maximize $(v - b)F^{N-1}[\beta^{-1}(b)]$.

This implies the first-order conditions

$$-F^{N-1}(\beta^{-1}(b)) + (N-1)(v-b)f(\beta^{-1}(b))F^{N-2}(\beta^{-1}(b))\beta^{-1\prime}(b) = 0.$$

Since $\beta'(\beta^{-1}(b)) = 1/\beta'(v)$, where $b = \beta(v)$, the first-order condition can be rewritten as

$$-F^{N-1}(v) + (N-1)(v-b)F^{N-2}(v)f(v)/\beta'(v) = 0$$

or

$$\frac{d}{dv}[(v-\beta(v))F^{N-1}(v)] = F^{N-1}(v).$$

Integration implies that

$$\beta(v) = v - \frac{K + \left\{\int_0^v F^{N-1}(s)ds\right\}}{F^{N-1}(v)}. \tag{4.1}$$

Note that the condition that a zero-value bidder gets zero implies $K = 0$.

It is relatively straightforward to calculate optimal bids when values are uniformly distributed. Let $[v, V]$ denote the interval over which values are distributed, and $L = V - v$. Then the expected value of the high bidder's valuation is $[N/(N+1)]L+v$, and this bidder's expected price is $[(N-1)/(N+1)]L+v$, which can be derived from (4.1).

The simplest case is when there are two bidders whose values are uniformly distributed over some interval, say [0, M]. In this case a bidder with value v will wish to choose its bid, b, to maximize $(v-b)(b/v)$. This leads to a solution $b=v/2$.

To calculate optimal bid functions more generally using this direct approach requires an explicit formula for the distribution function $F(v)$. The indirect RET approach provides a shortcut for calculating the amount of the second highest bidder's valuation. An exact amount may be no easier to obtain than when using the direct approach, but because one can approximately estimate the value for the second highest bidder, it can be easier to determine how to bid.

These approaches, however, assume that bidders are *risk neutral*. Consider a bidder with value v, who can offer either b or $\tilde{b}>b$. Let ρ be the probability of winning with an offer of b, and $\tilde{\rho}>\rho$ be the probability of winning with an offer of $\tilde{b}$. The expected payoff from bid b is is $E(b) = \rho(v - b)$, and from bid $\tilde{b}$ is $E(\tilde{b})=\tilde{\rho}(v-\tilde{v})$. In words, the offer of $\tilde{b}$ results in a lower payoff, with a difference $\tilde{b}-b$, but a higher probability of winning, $\tilde{\rho}-\rho$. This bidder can strictly prefer $\tilde{b}$ to b, even when $E(b)=E(\tilde{b})$, and vice versa, depending on the bidder's preferences toward risk. If the bidder is more concerned about winning than about the expected payment, it would prefer to bid more than an amount that maximizes $E(b)$.

4.3 Optimal Auctions

This section summarizes the key results from Myerson's famous (1981) paper on optimal auction design. An optimal auction, that is, one that maximizes expected revenues for the seller, can be derived using two key insights from Myerson's paper. The first insight is that for any auction there is an equivalent auction in which bidders are asked to report values. The second is that the optimal bid strategy determines the shape of the function for the payoff based on the reported type, up to an additive constant. The constant can be set to make the lowest value bidder indifferent between participating and not participating in the auction. This fully defines the optimal auction.

4.3.1 The Revelation Principle

Auctions for a single object can take many forms, and both losers and winners can be required to make payments. The following assumes, as above, that each bidder i's type, or value, v_i, is distributed according to a probability density function, $f_i(v_i)$. The values are distributed over some range $[a_i, b_i]$ for $i = 1, 2, \ldots, N$. A strategy for a bidder then is a choice of some action, from a set of permitted actions defined by the rules, namely $\alpha_i = \alpha_i(v_i) \in A_i$.

An auction "mechanism" is defined by the strategies and the rules for determining winners and payoffs. A strategy, in the context of an auction mechanism, is a choice of a bid amount as a function of the bidder's type, v. The winners are generally the highest value bidder, or bidders—in a multi-object auction. And a bidder's *payoff* is the value of winning (assuming it wins) less the payment it has to make.

A *direct revelation mechanism* is any allocation mechanism, or auction, in which bidders report values; the probability of a bidder winning, $\rho(\mathbf{v})$, and its payment, $x(\mathbf{v})$, will be a function of the vector of reported types of all the bidders, $\mathbf{v}$. Note that the actual type can differ from the reported type.

The revelation principle states that given any feasible auction mechanism, including ones in which bids do not represent values, there is an equivalent direct revelation mechanism that gives the seller and all bidders the same expected utilities as the given mechanism. The reason that the revelation principle holds is straightforward. Suppose bidder j of type v_j in an auction would have chosen action $a(v_j)$. Then an allocation rule that gives each bidder the probability of winning $p_j = p_j[a_1(v_1), a_2(v_2), \ldots, a_n)v_n)]$ and allocation of $x_j = x_j[a_1(v_1), a_2(v_2), \ldots, a_n)v_n)]$ would provide each bidder the same incentives and outcomes as the given auction. In other words, one can back out strategies bidders would choose to determine prices and allocations consistent with those that would result from bidders' optimal choices in the original auction.

4.3.2 The Revelation Principle and Optimal Auctions

The revelation principle was derived as a lemma in Myerson's 1981 paper and used in solving for optimal auctions. The payment rule and the rule determining the probability a bidder of a given type wins determine the bidders' behavior and the auction's outcome. This follows because the payment rule and the probability of winning determine bidding incentives. A bidder will prefer to report being of type v to any other type $v + \Delta v$ when $S(v) \geq S(v + \Delta v) - \Delta v P[(v + \Delta v)]$, where $S(v)$ is the net surplus a bidder of type v receives if it wins. As shown above, this condition fully defines the surplus function, and therefore the payment rule, up to an additive constant, $S_j(\underline{v})$, which in turn determines the payment of the lowest value bidder:

$$S_j(v) = S_j(\underline{v}) + \int_{\underline{v}}^{v} P_j(x)dx.$$

Now consider any two mechanisms that have the same $S_j(v)$ and the same $P_j(v)$ for all v and for all players j. In this case j's expected payoff is the same for both mechanisms. This means that the auctioneer gets the same expected revenues.

Therefore the term $S_j(\underline{v})$ determines the expected payoff to the lowest value bidder that actually bids. Variations in $S_j(\underline{v})$, through bidder participation fees, reservation prices, and the like, will affect the total auction revenues. Thus the auction revenues can be written as a function of $R = R(S_j(\underline{v}))$. The optimal auction, that is, the one that maximizes seller's expected revenues, corresponds to the choice of an expected surplus for the lowest value bidder. This in turn is equivalent to the choice of a reservation price.

To see what this value is, consider the case where the seller does not have any value for keeping the object, and where the bidders have values uniformly distributed on $[a, b]$. An optimal auction will be a modified Vickrey auction in which the seller sets a reservation price.

This is no longer a Vickrey auction, as the seller will have a preference to set a minimum price for the case where there is only one high bidder. The simplest case is that of one bidder—the seller will want to choose a reservation price of $(b-a)/2$. This means that the seller will sell the object with probability 0.5. Another way to see this is to look at expected revenue as a function of the reservation price r. This will be $r[1 - F(r)]$, and it is maximized at $r = (b-a)/2$.

The following calculations illustrate why the auctioneer will want to set a positive reservation price and risk the object not being sold. Suppose that the values are uniformly distributed over [0, 1]. When there is no reservation price and one bidder, the equilibrium in the totally noncompetitive "auction" is for this bidder to bid a bit more than 0 in a first-price auction, and its value in a second-price auction. This provides the auctioneer with zero expected revenues. Now, if the auctioneer sets a reservation price, its expected revenues are the reservation price times the probability that this bidder has a value above that level. The optimal reservation price is $\frac{1}{2}$.

When there are two bidders whose valuations are uniform draws from the interval [0, 1] and no reservation price, the expected revenue in a second-price auction is $\frac{1}{3}$. In contrast, when the auctioneer sets a reservation price of $\frac{1}{2}$, it will receive an expected revenue of $\frac{5}{12}$.[1]

More generally, for any number of bidders, n, the seller will want to choose this same reservation price at $(b-a)/2$ when bidders have uniform values on an interval $[a, b]$. The probability of the object not being sold is $\frac{1}{2}^{n}$. For a large number of bidders this probability gets arbitrarily small. When the distribution of values for different bidders can have different upper and lower bounds, it is possible that the revenue-maximizing auction will discriminate by allowing bidder-specific reservation prices, so that a bidder who has a higher value, when ex ante that bidder *is* expected to have a higher value, may still lose to a bidder with a lower value.

Also notice that an auction that maximizes expected revenues will not be efficient. When an object has no intrinsic value for the seller, then the efficient outcome is for the seller to award the object to the bidder with the highest value for it. An open first- or second-price auction with no reservation will be efficient in this sense. However, this will not maximize the seller's expected revenues. The seller will want to impose a positive reservation price.

When a government agency is auctioning off a resource, or concession, such as a spectrum license, oil lease, or highway concession, the optimal auction will not usually be the one that maximizes revenues. The agency may have other objectives, such as ensuring that the object is acquired and used. For instance, if an agency wants to promote development of the communications sector, it is likely to prefer allocating spectrum for free, if no one is willing to pay a positive price, with some sort of build-out requirement. This suggests that a government agency should set a zero reservation—unless the revenues lost from the optimal auction would need to be made up from a different source, such as increases in tax rates. Indeed there are times when an agency will set a negative reservation price; for example, in allocating spectrum or telecommunications concessions agencies may want to provide subsidies to ensure target coverage goals are met.

This analysis assumes all bidders show up. If bidder participation can be random or if there are participation costs, then another rationale for setting a positive reservation is that the auctioneer would have the possibility of withdrawing an object from the auction and re-auctioning at a later date. When participation at each auction date is random, then, as explained in more detail below in the analysis of multi-unit auctions, a positive reservation price could increase revenues and efficiency. This is very clearly the case when the auctions are conducted for future delivery of an object.

4.4 Incentive-Compatible Auctions

This section describes auctions that provide each bidder an incentive to report true values. The simplest form of such an auction mechanism is the second-price auction for a single object. For single-object auctions, the second-price sealed-bid auction is a mechanism under which it is a dominant strategy for bidders to report true values, that is, the auction is *incentive compatible*. What is key to this incentive compatibility is that a bidder's increasing its offer has no effect on the price the bidder pays *except* when this increase affects whether the bidder would have won, that is, when the bidder is *pivotal*. What is shown below is that this idea can be extended to multi-object auctions.

4.4.1 The Vickrey–Clark–Groves Mechanism

Recall from the revenue equivalence theorem (RET) that a second-price and a first-price auction for a single object will, with independent private values, result in the same expected outcome. However, it is *not* the case that bidder behavior will be the same. Vickrey noted that in a second-price auction each bidder has a dominant strategy to truthfully report the object's value. In contrast, in a first-price auction, each bidder will pay its bid, and each bidder will offer an amount less than its value. In equilibrium, the amount by which the winning bidder will shade its bid is the expectation of the second highest value. This section shows that in more general settings with multiple objects it is possible to fully characterize all auctions in which bidders will always want to bid truthfully. Such auctions are called Vickrey–Clark–Groves (VCG) auction mechanisms. A second-price auction is just a special case of a VCG mechanism.

What follows assumes *additively separable* utilities, that is, a bidder's utility of winning an object or allocation x can be written as $u_i = v_i(x) + t_i$, where t_i is the monetary transfer that i receives. The revelation principle states that it suffices to consider only direct revelation mechanisms in characterizing auctions in which bidders will have incentives to truthfully report valuations. In other words, bids can be assumed to take the form of reported valuations for an outcome or an allocation.[2]

A VCG auction is a mechanism by which a bidder's transfer satisfies the following condition:

$$t_i[\mathbf{w}(\mathbf{x})] = \sum \mathbf{w}_{-i}[\mathbf{x}^*(\mathbf{w})] + \mathbf{h}_i(\mathbf{w}_{-i}(\mathbf{x})), \tag{4.2}$$

where $x^*(\mathbf{w})$ is the allocation when $\mathbf{w}(\cdot) = (w_1(\cdot), \ldots, w_n(\cdot))$ is the vector of the reported valuations and $\mathbf{w}_{-i}$ is the vector of valuations of all bidders other than i.

Note that i's bid only affects its transfer, or payment, in (4.2) to the extent that i's bid changes the allocation that other bidders receive. If i's bid is not pivotal, then i's bid has no effect on its monetary payment. Further it can be shown that a mechanism that satisfies (4.2) has the property that i's transfer will equal the amount by which i's bid affects the aggregate welfare of the other bidders, that is,

$$t_i(\mathbf{w}_{-i}, w_i) - t_i(\mathbf{w}_{-i}, w_i') = \sum \mathbf{w}_{-i}(\mathbf{x}^*) - \sum \mathbf{w}_{-i}(\mathbf{x}'), \tag{4.3}$$

where x^* is the allocation at (w_{-i}, w_i) and x' is the allocation when the reported valuations are (w_{-i}, w_i'). In other words, i's report only affects i's transfer to the extent that this report affects the aggregate welfare of other bidders. Indeed this latter condition is essentially a necessary and sufficient condition for an auction mechanism to provide bidders an incentive to bid true values.[3]

4.4.2 Properties of VCG Auctions

The second-price auction and, more generally, the VCG auction has the desirable property of eliciting truthful revelation of preferences and of achieving efficiency. This raises the question why the VCG auction should not be used more generally. This section addresses this issue.[4]

One property of VCG auctions is that a winning bidder does not have to pay its bid amount. The gap between the winning bid and the payment can be quite large. Auctions in which the second price was less than 2 percent of the winning bid occurred among the first spectrum auctions.[5] This gap does not mean that there was necessarily any lost revenue. However, it does suggest that a reservation price or a higher reservation price might have been advisable. Such large gaps can look bad for the auctioneer and also require post-auction rationalization, they invite challenges and litigation, and they may result in cancellation or re-auction. A first-price auction or an English auction avoids this second-guessing. And, in some situations, an English auction will also result in higher revenues.

VCG and Budget or Revenue Constraints The VCG auction has a number of other properties that may discourage its adoption. Recall that in a VCG auction, a bidder payment is given by (4.2). One condition that might be needed to ensure feasibility of the auction is some budget balance requirement. Or, in a government-managed auction, the revenues might be capped or required to be distributed back to the participants net of administrative costs. For example, the auction can include contributions by users and offers from suppliers. An agency conducting such an (two-sided) auction might not be allowed to incur a deficit. Absent some additional restrictions, the sum of the bidder payments in (4.2) can otherwise be arbitrary. This limitation on transfers to bidders imposes the condition that the sum of the terms in (4.2) is zero. However, as Milgrom (2004) has shown, this type of restriction can be incompatible with the requirement for a VCG auction to exist in (4.3).

To see this, consider the case of two bidders for one object.[6] The required condition is that there are functions h_1 and h_2 that always add up to zero Suppose that bidder 1's possible values for the object are either 1 or 3, and that bidder 2's possible values are either 2 or 4.

- If bidder 1's valuation is 1 and bidder 2's valuation is 2, then bidder 1 should pay $p_1 = h_1(2)$ and bidder 2 should pay $p_2 = 1 + h_2(1)$. So we seek $p_1 + p_2 = 0 = 1 + h_2(1) + h_1(2)$.
- If bidder 1's valuation is 3 and bidder 2's valuation is 4, then it must be the case that $p_1 + p_2 = 0 = 3 + h_2(3) + h_1(4)$.
- If bidder 1's valuation is 1 and bidder 2's valuation is 4, then it must be the case that $p_1 + p_2 = 0 = 1 + h_2(1) + h_1(4)$.
- If bidder 1's valuation is 3 and bidder 2's valuation is 2, then it must be the case that $p_1 + p_2 = 0 = 2 + h_2(3) + h_1(2)$.

Adding of the first two and last two equations from the list above requires that

$$0 = 4 + h_2(1) + h_1(2) + h_2(3) + h_1(4) = 3 + h_2(1) + h_1(2) + h_2(3) + h_1. \quad (4.4)$$

This is impossible, so the budget balance fails. More generally, this means that a VCG auction may fail to exist when the auction must also meet specific revenue targets, even when the revenue targets are consistent with bidder valuations, such as raising zero net revenues in the discussion above.

Other VCG Auction Properties In both single object and multi-object VCG auctions the price paid by any winning bidder is determined by rivals' losing bids. Most of the analysis in auction theory, and in this

Primer, assumes bidders do not care about what rivals pay. In practice, this is not always the case.[7] Managers can be judged by relative performance.

Also bidders' true values may be information of interest post-auction, to the seller and to other parties. If there is any chance that bid data will be used to set fees, prices, or other costs for a bidder post-auction, then bidders will have disincentives to bid truthfully.

4.4.3 Multi-Object VCG Auctions

This section provides a brief description of some multi-object VCG auctions; chapter 10 provides more detailed analysis of multi-object VCG auctions. Where bidders are each seeking to purchase at most one object, the second-price auction can be readily modified for the case of k identical units of the same product. More specifically, when all units are sold to the k highest bidders, but for the $k + 1$st highest price, then each bidder will have as a dominant strategy to bid its true value, as is the case in the standard second-price auction.

Several spectrum auctions have been conducted using a modified VCG format.[8] However, for the most part, the modifications leave bidders with some incentives to deviate from truthful bidding. One variation of $k + 1$st-price rule was used for selling identical spectrum licenses in Denmark in 2011.[9] That auction included four generic licenses, and bidders could win one license each. The four highest bidders won but paid the fourth highest bid amount. In this case it was no longer true that a bidder would always want to bid its value. When it is possible for a bidder to be a marginal bidder, then the bidder would want to shade its offer, just as in a standard first-price auction.[10]

Table 4.1
New Zealand 8 MHz UHF TV license auction

Lot	Winner	High bid	Next bid
1	Sky Network TV	$2,371,000	$401,000
2	Sky Network TV	$2,273,000	$401,000
3	Sky Network TV	$2,273,000	$401,000
4	BCL	$255,124	$200,000
5	Sky Network TV	$1,121,000	$401,000
6	Totalisator AB	$401,000	$100,000
7	United Christian	$685,200	$401,000

Another series of spectrum auctions in New Zealand used a second-price rule (Mueller 1993). However, these auctions were not incentive compatible, as bids for each identical license were taken separately, and the price for each license was the second highest bid for that license rather than the highest losing bid for all seven licenses, meaning the eighth highest offer. Table 4.1 from Mueller (1993) shows one such *simultaneous* auction for seven identical licenses. The prices paid ranged from NZ$100,000 to NZ$401,000.

The New Zealand government also conducted a set of *separate* second-price auctions for three cellular licenses. The relevant bids for these three auctions were as follows (tables 4.2, 4.3, and 4.4 show the results):

Table 4.2
New Zealand AMPS A auction

Bidder	Bid amount
Telecom New Zealand	NZ$101,200,000
First City Capital	NZ$11,158,800
Imagineering Telecommunications	NZ$1,388,000

Table 4.3
New Zealand TACS A auction

Bidder	Bid amount
Bell South	NZ$85,552,101
Telecom New Zealand	NZ$25,200,000
Racal-Vodafone Ltd.	NZ$1,000,000
Broadcast Communications Ltd.	NZ$2,000

Table 4.4
New Zealand TACS B auction

Bidder	Bid amount
Bell South*	NZ$85,552,101
OTC International*	NZ$13,250,000
Telecom New Zealand*	NZ$7,000,000
Broadcast Communications Ltd.	NZ$5,000
Michael Oliver Thaisen	NZ$300

The winning bidders had to pay NZ$11,158,800 and NZ$25,200,000 for the AMPS-A and TACS-A licenses. The three high bidders for the TACS-B license (marked by an *) were disqualified, and the entire auction was rerun. In these auctions for cellular licenses, the bidders had to determine how to bid across auctions. Even where the bids are collected at the same time, bidders still need to guess how to bid.[11]

As noted above, there are incentive-compatible VCG designs for multi-object auctions. Chapter 10 characterizes prices and outcomes in such auctions, as well as the variants of VCG auctions being used for selling spectrum licenses in a number of European countries.

4.4.4 VCG-like Mechanisms for Advertising Auctions

One area in which a variant of the second-price auction is commonly used is for position in the ad sections of online search engine web pages, such as Bing and Google. When someone enters a keyword, for example, "plumber," the top of the screen will contain a small number of ads. Which ads appear depends on continuous online bidding for the positions. These auctions now generate billions of dollars in annual revenues.[12] The following describes the mechanism used for these auctions, which is a variant of the VCG auction.

To use Varian's (2007) notation, it is assumed that there are A agents competing for $S < A$ slots. The bids typically take the form of a fee paid by the winner per *click-through*. The higher slots are assumed to have a higher *click-through rate* (ctr), x_s. So a bidder that derives a profit of v_a per click will earn a profit of $u_{as} = v_a x_s$ from winning slot s. The slots are numbered in such a way that $x_j > x_{j+1}$ for all $j = 1, 2, \ldots, S - 1$. Table 4.5 summarizes the Google auction.

Bids for these slots take the form of a bid per click-through. So, if a bidder with value per click-through v_a pays $p_s = b_{s+1}$ for slot s, then that

Table 4.5
Position auction bids and payoffs

Position	Value	Bid	Price	CTR
1	v_1	b_1	$p_1 = b_2$	x_1
2	v_2	b_2	$p_2 = b_3$	x_2
3	v_3	b_3	$p_2 = b_3$	x_3
4	v_4	b_4	$p_1 = b_4$	x_4
5	v_5	b_5	$p_1 = 0$	0

bidder's profits are $x_s(v_a - p_s) = x_s(v_a - b_{s+1})$. In other words, the bidder that wins slot s pays the value the next highest bidder would have derived from that slot.

Varian shows that this pricing rule will result in the same allocation and payments at a symmetric Nash equilibrium as the VCG pivot mechanism.[13] Thus the position auction does provide bidders some incentives to bid truthfully. However, this result rests on a number of assumptions. I briefly list a few.

First, it is assumed that bidders each would receive the same ctr, x_s, from a given slot. It is unclear why this assumption is reasonable. Indeed bidders may place different priorities on getting the top slot.

Second, bidders may have advertising budgets limiting what they can bid. Suppose, for example, that there are two bidders, each having a budget of 100, and each wanting to buy a high position for the same three keywords. There will be no way for two bidders to allocate their budgets so that the resulting outcome is a Nash equilibrium in pure strategies. Each will want to try to pick up one slot cheaply, leaving a lot over for a second slot. Indeed there is no equilibrium in pure strategies.

Third, values of some slots can depend on other slots won. There can be both substitutes and complements. The Google auction does not allow for the type of contingent or package bids that are required if bidders are to convey true values.

4.5 Summary

This section concludes with a brief discussion and example of how the principles of optimal auction design discussed in this chapter can be put into practice. As is explained below, there may not be a single optimal auction design, or if there is one, the conditions under which it applies may be overly restrictive. This discussion uses an example from an actual auction design problem to illustrate some of the main issues.[14]

The design problem being addressed was to determine how to best allocate universal funds to support universal telecommunications service goals in the United States. The US Federal Communications Commission (FCC) and the state regulatory commissions established the Joint Board on Universal Service. The purpose of this agency is "to make recommendations to implement the universal service provisions of the Act. This Joint Board is comprised of FCC Commissioners, State

Utility Commissioners, and a consumer advocate representative."[15] The Act established a fund to promote deployment of telecommunications services in rural areas in for underserved segments.

The problem was to find an efficient auction design. This design had to meet the objective of minimizing costs for each market area receiving funding. In other words, the auction design was to maximize the benefits that could be provided with a fixed available amount of universal service funds. The auction design needed to take account of a number of features of the market. First, there was the problem of dividing the market into areas of a size appropriate to the potential service providers. This is a problem addressed in more detail below. For universal service funds, the potential service providers included wireline telephone companies, cable television, and cellular firms. This suggested certain geographic areas.

One additional concern was that there could be post-auction holdup or performance failures. Auction winners would be more likely to meet their universal service commitments if there was competition in the market. Unfortunately, this form of competition can result in excessive costs due to unnecessary duplication of facilities. The proposed way to address this trade-off was to include in the benefits accruing to consumers an extra factor for the presence of post-auction competition in a market. This meant that a second provider would be selected when the costs imposed by that provider were below a certain level.

The auction design problem was to derive conditions on the set of winners and payments that had to be satisfied at an optimum. What was shown was that an optimal auction design was any design for which the efficient set of providers almost always won, and for which the highest cost winners earned zero profits.[16] What this result does not provide is an explicit characterization of the optimal auction design; rather, it indicates than there is an entire class of auction designs that all achieve the same outcome. This general result does have some directly useful implications. For instance, it indicates that the market structure should be endogenous. However, it is left for the auction designer to develop specific rules and verify they do indeed satisfy the conditions for an optimum allocation.

5 Imperfect Information and the Winner's Curse

Summary

This chapter describes auctions in which bidders have imperfect information about the value of the object for which they are bidding. One specific case is that where the object has the same common value for each bidder, but bidders have different ex ante estimates of the value. This chapter also examines auctions in which bidders have affiliated values; that is, bidders' values are correlated in some way. What this means is that winning provides information, and so bids should be adjusted to allow for this fact.

5.1 Common Values and Order Statistics

Competing bidders often have to bid without having perfect certainty over values. A bidder may have an ex ante belief, or prior, about the value. That prior, which can be modeled as a probability distribution, and one in which the possible range of values is bounded both above and below, may be correlated across bidders. For example, in any auction for a durable asset with a resale market, such as real estate or a building, bidders' values will be correlated, as part of each bidder's value is the potential gain from resale. In the *pure common-value* case, all bidders will have the same value ex post, even if they have different forecasts of value ex ante. Bidders for oil leases will usually have only imprecise, and sometimes asymmetric, information about the amount of oil available in the tract being auctioned,[1] but all bidders may have the exact same value ex post—the value of the oil in that tract. In any auction in which value depends on future prices, such as oil for oil leases, or on complementary products, the bidders' valuations must contain some common-value components.

In a common-value auction the winning bidder will tend to have the most optimistic forecast value, and thus tend to overestimate the value. This will be true, for instance, when each bidder has access to a signal, or forecast, of the value of an object; then the bidder with the highest signal will generally want to bid the most. In this case the winning bidder will tend to overpay whenever the underlying value is the same for all bidders, as in a pure common-value auction, unless that bidder adjusts its bids to allow for the fact that being high bidder means it has the most optimistic signal. If each bidder bids up to its ex ante, unconditional expected value, as it would in a second-price auction in the pure private values case, the winning bidder will likely bid too much.

To see why this is the case, suppose that the true value of an object is V, and n bidders have signals s_j, $j = 1, 2, \ldots, n$, uniformly distributed on $\left[V-\frac{k}{2}, V+\frac{k}{2}\right]$. If bidder j offers s_j, then the expected price is $(V+k(n-3))/2(n+1)$, which is greater than V whenever there are four or more bidders. In other words, the bidder with the highest signal will have a value that, on average, will exceed the true value by $k[(n-1)/2(n+1)]$. Knowing this, such a bidder should discount its bid accordingly. A naive bidder, failing to make such an adjustment, might win the auction, but suffer losses. Such was the experience of bidders for oil leases as documented in Capen, Clapp, and Campbell (1971).

It is easy to see that the RET can fail when there are common values, or even when bidder valuations are affiliated, that is, imperfectly correlated. Consider the above example of a common-value auction. In a common-value auction, the rival bidders can have accurate forecasts on average, but the winner in each auction will tend to be that bidder that has the most optimistic forecast in that auction. In the second-price auction, bidding the ex ante signal of its valuation is the dominant strategy when values are independent. However, while it still might be the case that a bidder should bid its ex ante value in a second-price auction, conditional on being the winner among n bidders, it is no longer true that this value should be the stopping price in the English auction is adjusted each time a different bidder drops out. The question addressed here is how auction formats compare with common or affiliated values.

This chapter also addresses the more general question of how imperfect and asymmetric information can affect bidding behavior and the auction outcome. More specifically, the precise information structure can affect bidder behavior. When bidders have affiliated values, the structure of information can matter a great deal. As is explained in further detail below, it can matter more whether information is private than whether it is precise.

5.2 Bidding Strategy in Common-Value Auctions

This section assumes a pure common-value auction, that is, each bidder i has a signal $x_i = v + \varepsilon_i$ of the true value v.[2] It is also assumed that the signals are drawn from a common density function $f(\varepsilon)$. Note that the expected value, conditional on all n bidders' signals, is $\sum x_j/n$. While it is true that $E(x_j) = v = \sum x_j/n$ (where $E(Z)$ denotes the expected value of Z), it is not the case that $E(X^1) \equiv E(\max_j\{x_j\}) = v$. In other words, the likely winning bidder will usually be the bidder that has the highest signal and will have an expected value, conditional on winning, that does not equal the unconditioned expected value. Thus, in an auction, the bid strategy must be adjusted for this bias, to avoid the *winner's curse*. This adjustment will be based on the information available to the winning bidder, and that bidder has to make his or her final decision about a last and best offer. Because the information available can depend on the type of auction, the expected revenues of different auctions, such as English and Dutch, need not be the same.

5.2.1 Bid Strategy in Dutch and First-Price Auctions

In a first-price auction, (4.1) determines the optimal bidding strategy, assuming independent private values. As above, let $\beta(s)$ denote the equilibrium bid function for all other bidders, and assume a symmetric equilibrium. In the case of common values, it is still true that a bidder with a signal s of value v will, at equilibrium, not want to shade its bid below (increase its bid above) $\beta(s_i)$ when the lower (the higher) probability of winning will not exceed the benefits of reducing (increasing) its cost should it win. That is, using the notation of chapter 4,

$$E\{\mathrm{P}(\beta(s_i + \Delta s))[v - \beta(s_i + \Delta s)] \mid s_i + \Delta s \geq s_j, j \neq i\} \leq E\{\mathrm{P(b)}v - \beta(s_i) \mid s_i \geq s_j, j \neq \mathrm{i}\},$$

where, as above, $P(b)$ is the probability the bid b wins.

This condition just states that a bidder with signal s_i will not want to report it has a different signal $s_j \pm \Delta$. Moreover this condition is the same as used in deriving (4.1) except that the expectation was not conditional on the bidder having the highest value signal. More to the point, in the independent private-values case of chapter 4, the probability that a bidder would attach to winning is not affected by others' values, whereas bidders' values are so affected in the common-value model. Thus it is still the case that the optimal bidding rule has the same form as in (4.1), with the exception that the distribution function is conditional on the bidder in question having the highest value, that is,

$$\beta(v) = v - \frac{K + \left\{ \int_0^v F^{N-1}(s \mid v > s) ds \right\}}{F^{N-1}(v \mid v > s_j \text{ for all other bidders } j)}. \tag{5.1}$$

It is also the case that a bidder in a first-price auction will face exactly the same trade-offs as in a Dutch auction. In both cases each bidder will want to shade its bid a bit, that is, bid below value. Moreover in each case a bidder will face the same trade-offs in making that decision. A winning bidder will want to enter a bid, at an amount less than value, that maximizes its expected surplus conditional on its having the most optimistic value. Thus the strategic equivalence of the Dutch and first-price sealed-bid auctions carries over to the common-value auction.

5.2.2 Bid Strategy in a Second-Price Auction

In a pure common-value auction the firm with the highest value estimate will win and pay the amount offered by the bidder with the second highest value. So the bidder that wins will have the highest signal, and its offer should be discounted accordingly. The offer should equal how much the bidder expects the value to be, conditional on its forecast having the highest value. The amount that a bidder should offer is still equal to its true *expected* value. This offer should be based on the assumption that, in the worst case, it will be tied for having the highest forecast with one other bidder. Moreover the expected value is no longer the signal or forecast value.

5.2.3 Bid Strategy in English Auctions

When bidders' valuations are independent, the optimal bid strategies in the second-price and English auctions are essentially equivalent. In the English auction, a bidder will want to bid up to its forecast value, and then drop out; in the second-price auction, a bidder will bid its ex ante forecast value. The outcome is also the same. To see this is no longer true when there are common values, it suffices to consider the decision facing a bidder about when to drop out in an English auction.

Consider a bidder that has a signal s about the value of the object being auctioned. Further assume that this is a pure common-value auction. What follows describes the bidding decision for a Japanese-style English auction, that is, an auction that starts with a low price that the auctioneer gradually increases until there is only one bidder remaining. Bidders, at each increment, must indicate that they are still in the auction or drop out. Once a bidder drops out, it is never allowed back. And bidders each see how many bidders remain after each bid increment.

If the bidder has not seen any rival drop out, then the worst case—that is, the case where the bidder would have the lowest conditional forecast value—is when all rivals have a signal that is just the same as its own. So, if the true mean forecast is $\sum_{k=1}^{n} s_k/n$, a firm with the signal s should drop when the price reaches the expected value $E[V \mid s_k = s \text{ for all } k]$.

Let $p^1(s)$ denote the value of the first drop. Notice that this is a strictly lower amount than a bidder might forecast to be the expected value when that bidder assumes it does have the lowest value signal.[3]

The next drop must occur when the bidder with the second lowest value signal can expect to lose money if it should stay in. The decision about when to drop for the bidder with the second lowest value signal needs to be based on what its forecast would be, conditional on its winning. The only way such a bidder (call it bidder 2) can win is if all remaining rivals have the same value as it does, unless it bids too much. Bidder 2 will bid too much if it stops after higher value bidders have dropped out. The worst case for bidder 2 occurs when every rival has the same value, just a bit more than bidder 2's value, and they all drop out at a price just above where bidder 2 might. Bidder 2 can wait, with the expectation that the actual value is better than the worst case; however, this bidder does not want to win if it actually has the second lowest signal, and so its choosing to drop out, assuming the worst case, is, in this instance, a dominant strategy.[4] Let $\beta_k(s, h_{k-1})$ denote the strategy for the kth bidder to drop out given that the history of the $k-1$ previous drops is h_k and the bidder has signal s. Also suppose that bidders are indexed in order of the value of their signals.

Let $\beta_2(s, p^1) = \mathrm{E}[V \mid \beta_1(s) = p^1(s), s^j = s \text{ for all } j \geq 2]$, where s^j is the jth highest signal. Then $\beta_2(s,p^1)$ specifies the optimal price at which to drop out of an English auction for a bidder with the second lowest signal. One can iteratively define an optimal stopping price for each bidder,

$$\beta_k(s, h_{k-1}) = E[V \mid s^j = s \text{ for all } j \geq k, h_k].$$

These strategies yield optimal prices for a bidder to drop out, except for the last two bidders. The last two bidders will see that all other bidders have dropped out. Thus, the remaining bidder with the lower signal should assume its last remaining rival has a higher signal. For example, suppose that there are two bidders to start the auction. Also, suppose that these two bidders have signals that are $s^j = V + \varepsilon_j$, where ε_j is uniformly distributed on $\left[-\frac{1}{2}, +\frac{1}{2}\right]$. If the bidders do not know the true value V, then the lower value bidder should drop out when the price reaches its signal plus $\frac{1}{6}$, as the expected value of the lower value signal will be $V - \frac{1}{6}$, and that of the higher value signal will be $V + \frac{1}{6}$.

This induction approach defines optimal bid strategies for a sequence of common-value English auctions. Risk-neutral bidders will drop out in the order of the value of their signals, and the bidder with the highest signal will win, but will only pay the price at which the second-highest-value bidder drops out.

5.2.4 Comparisons of English, Second-Price, and Dutch Common-Value Auction Outcomes

The preceding discussion provides an analysis of the optimal bidding rules in English, Dutch or first-price, and second-price auctions when there are common values. Common values mean that bidders need to adjust their bids to reflect that winning provides information. In other words, the winner will tend to have the most, and an overly, optimistic forecast.[5] In many auctions the bidders' values have common-value components. Capen, Clapp, and Campbell (1971) observed that bidders for oil leases often failed to adjust for the winner's curse, and as a result winning bidders tended to overbid. These were not pure common-value auctions, in that bidder values included not only the common value of the underlying oil but also the knowledge, costs, and ability to exploit the oil lease, which could differ across bidders. However, bidders still needed to adjust for the likelihood that winning means that a bidder has made an overly optimistic forecast.

As an illustrative example, suppose bidder signals of values are distributed uniformly on some interval $\left[V - \frac{1}{2}, V + \frac{1}{2}\right]$, and bidders do not know V. Suppose that the true value is V. So the expected value of the jth lowest signal will be

$$s^j = V - \frac{1}{2} + \frac{j}{n+1}.$$

In contrast, in an English auction, the bidder with the jth lowest signal will want to bid up to the point at which the price equals

$(s^1+s^j)/2$. The expected value of this price is then $V/2-V/2(n+1)$. That is, in the English auction the expected price is higher and closer to the expected value. Although this is not always true, when there are common values, the expected price in the English auction cannot be less than in the second-price sealed-bid auction.

And finally, in a first-price sealed-bid auction, a bidder with signal s that offers $\beta(s) = s - x$ will have a probability of winning of $s-V+\frac{1}{2}$, and will win $V-(s-x)$. The bidder will want to maximize the expected value of this with respect to x. It can be shown[6] that the optimal bid is $\beta(s)=s-\frac{1}{2}$, or $x=\frac{1}{2}$. This results in an expected value for the price of $V-[1/(n+1)]$.

Summarizing, the expected price is as follows:

- In an English auction: $V-[1/2(n+1)]$.
- In a second-price auction: $V-[(n-1)/n(n+1)]$.
- In a first-price auction: $V-[1/(n+1)]$.

So in this example, and more generally, the equilibrium price in an English auction will exceed the price from a second-price auction, which in turn will exceed the price from a first-price auction.

5.3 Almost Common-Value Auctions and Informational Asymmetries

Bidders can differ in the amounts of information they have access to and in their underlying values. For example, in the types of oil lease auctions described in Capen, Clapp, and Campbell (1971), some bidders may have access to better geological information than their rivals. Moreover there are investments in complementary facilities that some bidders may have already incurred and rivals have not.[7] The optimal auction strategy, and thus the outcome, will depend a great deal on how information is distributed among the bidders.

The preceding section considered the case where, ex ante, all bidders have the same quality of information, although not usually identical information. Bidders may all derive value from the same unknown variable, such as the amount of oil in an oil tract, but will typically have different views or forecasts as to the magnitude of this common value. This type of model is sometimes referred to as a mineral rights or an oil lease model.

However, bidders need not be symmetric. First, some bidders may be better informed than their rivals. This is the situation in drainage or resuscitation tract auctions; in such auctions those bidders that have previously explored or developed the tract will have knowledge about reserves and costs that rivals do not have. Sometimes informed bidders may not all have access to the same information. In other cases bidders will have some private values, that is, economic advantages. This occurs, for instance, when some bidders have previously invested in complementary assets, such as storage and pipeline facilities. This cost advantage, or value premium, gives these bidders an advantage over rivals.

5.3.1 Auctions with One Informed Bidder

This subsection considers the situation in which one bidder is informed, and the other bidder or bidders are uninformed, about the value of the object for sale. In the simplest case, the value can be high, H, with ex ante probability ρ, and low, L, with probability $1 - \rho$. Assume that this is a first-price sealed-bid auction. In this situation the informed bidder will want to bid just high enough to make uninformed bidders indifferent between bidding and not bidding. More specifically, let $x = \rho zH + (1 - \rho L)$, where z is the probability that an uninformed bidders wins when the value is actually H. If the uninformed bidder always bids x, and the informed bidder offers x when the value is H, then the uninformed bidder will be indifferent between bidding and not. Any lower bid will mean that the uninformed bidder can bid a bit more and always win when the value is high. Any higher bid will mean the uninformed bidder would not bid.

This logic can be generalized for any number of uninformed bidders. Somewhat surprisingly, the outcome is the same. One key requirement for an equilibrium is that uninformed bidders must just break even, conditional on winning; otherwise, the informed bidder should adjust its bid. If there is only one informed bidder, then the equilibrium bid is the same no matter how many uninformed bidders there are. The other key requirement for an equilibrium is that no bidder has an ex ante negative expected payoff.

Notice that a pure common-value English auction ends up as essentially a first-price sealed-bid auction for the two bidders that have the most optimistic forecasts. After all but the last two bidders have dropped out, each of the two remaining bidders will know its own signal, and each will know that if it is not the last to drop, then it will have the highest value signal. The one with the high value signal will know this value, which its only remaining rival can only guess at.

Analytically, the firm with the lower forecast of the two will not want to wait too long to drop. The difference is that in the common-value auction, the last two bidders do not know which is the stronger until after the auction ends.

5.3.2 Auctions with Both Informed and Uninformed Bidders

When there are two or more informed bidders, and uninformed rivals, the uninformed rivals will generally be at a critical disadvantage. If the informed, or better-informed, bidders have identical information, then they will bid prices down to the point where uninformed bidders cannot break even. In other words, uninformed bidders do not want to win. For any common signal the informed bidders will never, in equilibrium, submit any offer that will leave a positive profit. If one informed bidder did submit such an offer, then an informed rival would seek to submit a slightly higher bid amount. Thus, if an uninformed bidder wins, it will have bid too much, and will lose money. This situation is then logically equivalent to a pure common-value auction among the informed bidders, with uninformed bidders not bidding.

The situation is different when bidders have different information. In that case, if several bidders have access to the same information, then none of those bidders will be able to earn positive expected profits.[8] However, if one bidder has access to even very imperfect signals about the object's value that its rivals do not have, then it can earn a profit. For example, suppose that one bidder has a signal of value that has a high variance, and several other bidders have access to an identical, low-variance signal. The bidders that have the same information will compete away profits. When the bidder with different information has a positive signal, while the others have a negative or lower signal, it can win and earn a positive return. In other cases it will not win.

5.3.3 Other Asymmetries

Most auctions have common-value aspects. Often bidders differ in many other ways. After any uncertainty is resolved, bidders can have much different underlying values—for instance, due to differences in costs or ability to derive benefits from what is available. If a single bidder has a value or cost advantage as well as an informational advantage, then the analysis is essentially the same as in the case where there is a single informed bidder competing with one or more uninformed bidders.

However, with several informed bidders, only one of which has a value or cost advantage over its rivals, the disadvantaged, but better informed bidders should not win; nevertheless, their presence will mean that uninformed bidders should not bid, or bid only very low. An uninformed bidder can lose despite having a value or cost advantage over an informed rival. This case is more complex in that the information advantage can fully or partially offset the benefits of the value differential.

More specifically, the informed bidder may choose to bid low, and accept a low or zero probability of winning, if it has a negative signal of the value, and to bid aggressively if it has a positive signal. The bidder with the higher value may want to reduce its bid, knowing it will derive lower net value if it wins. If it may be the case that the strong bidder will never want to allow the informed bidder to win: it will want to keep its bid high enough to win all the time. Alternatively, the strong bidder may want to allow the informed bidder to win when it has a positive signal. In this case the informed bidder will bid just high enough to top the informed bidder when the value is low.[9]

5.4 Summary

This chapter has examined the influence that uncertainty in a common or affiliated value has on the outcome of an auction. Bidders facing this type of uncertainty will want to adjust bids to offset the impact of the winner's curse: winning means a bidder has the most optimistic forecast, which will usually be too high. One of the key results of this chapter is that an open auction allows bidders to aggregate information. This reduces the winner's curse and increases auction revenues. Thus an English auction will generate greater revenues than a first-price or second-price sealed-bid auction.

This chapter also explains how one can calculate an optimal bid strategy to correct for the winner's curse. The calculation essentially involves correcting estimates of values and optimal bids for the fact that having the winning bid means that the bidder has the highest (most optimistic) signal. It turns out that the additional information available to the winning bidder in an English auction increases the precision of value forecasts. The second-price auction will also generate more revenue than a first-price auction.

This type of analysis has had some practical and observable implications for oil lease bids. This is one situation in which bidders have experienced the winner's curse, and presumably learned how to adjust their bids. The empirical evidence[10] suggests that bidder experience has allowed bidders to adjust bids in line with what is an ex ante optimal bid strategy. Green and Porter (1984) also suggest that this is also likely to be the case with asymmetric information.

Undoubtedly, the winner's curse arises in other auctions, but it is less well documented than it is for oil lease bids. There is, however, extensive experience with spectrum auctions, and there are numerous cases where a bidder has won only to find it overpaid. The most striking example is the German 3G auction in 2000. In that auction two bidders, Group 3G (a consortium that included Sonera and Telefonica) and Mobilcom Multimedia (with France Telecom as the major mobile operator in the consortium), each paid approximately $8 billion for a license, only to abandon it a short while after the auction. Neither operator ever launched.[11] These were very large, and presumably well-prepared and sophisticated, firms.

The ability of bidders to offset for the winner's curse can be very limited. Experimental evidence suggests that the calculations can be hard for bidders to make accurately. As the theory suggests,[12] providing more information to bidders should improve welfare and efficiency. In practice, however, it is unclear in many cases (especially one-off auctions or auctions that are very sporadic for bidders) how to correct accurately for the winner' curse.

6 Sequential Auctions of Substitutes

Summary

This chapter examines sequential auctions of multiple lots of an identical good or service. The question addressed is whether price systematically increases or decreases from one auction to the next. Such systematic price patterns can be an important consideration for bidding strategy and for auction design. First considered is a sequence of auctions with an identical pool of risk-neutral bidders. It is shown that under fairly general conditions the expected price will tend to neither increase or decrease across auctions. It is also shown that when bidder opportunity costs can change from one auction to the next, or when bidders are risk averse, definite price trends will arise.

6.1 Multi-Lot Auctions

A single auction or auction event will often include multiple units of identical or similar products. This chapter examines auctions in which the lots are sold sequentially, generally one at a time. Subsequent chapters consider alternative auction designs in which a single auction process, or even a single package bid, can include multiple units, or even multiple objects.

Bidders participating in a sequence of auctions will need to decide how high to bid in any given auction when there are other auctions to follow for identical or similar objects. This is especially the case in a sequence of auctions; an aggressive bid early in the sequence can lead to regret when prices are significantly lower in subsequent auctions. Conversely, a bidder who decides not to bid up to his or her value in an early auction can be disappointed with much worse prices, and the possibility of being shut out, in subsequent auctions.

This chapter describes solutions to the sequential decision problems facing the auction originator and the bidders. The analysis of the solutions to these decision problems is then applied to characterize expected equilibrium outcomes in different types of sequential auctions. Most of what follows assumes that each bidder is interested in at most one unit, so that strategic withholding is not a consideration. Thus a bidder considering whether to improve its offer for any lot will have to weigh the extra probability of winning the current auction against expected surplus that it will receive for waiting for one of the next lots. The number of remaining lots and the number of bidders can, in theory, affect this calculation, and in a surprising way. This calculation can be affected if bidders are risk averse, or if their opportunity costs change across auctions.

The auctioneer, at times, will also have a choice about how to allocate what is available for auction over time and across auctions. This type of timing decision is common in energy procurement[1] and in many other sectors.[2] As is explained below, this decision can affect overall revenues. In addition the auctioneer may consider different reservation prices, and different ways to split the quantity available across auctions. The auctioneer can consider reservation prices based on offers in the auction, choosing to defer some volume to the next if competition is low, and the reverse if competition is high, in an initial auction; in other words, the auctioneer can choose the quantity to carry over from one auction to the next, based on the level of competition as well as on other factors such as pre-auction price expectations.

These decisions become more complex when the products are imperfect substitutes. In this case the order of what is offered can affect bidder decisions and the overall outcome.

6.2 The Declining-Price Anomaly

An item sold in one auction will often have a copy, or a similar item, available in subsequent auctions. It has been observed that prices will tend to vary quite a bit even when identical objects are auctioned one after another in the same auction event. This suggests that auction design and strategy can affect the outcome. This also appears to contradict one of the fundamental results of economic theory—that all parties face the same price for the identical product in a competitive market. Ashenfelter (1989) refers to this lack of price uniformity in

Table 6.1
Sequential wine auctions: distribution of price patterns for identical wines sold in same auctions (number of auctions)

	Christie's London	Sotheby's London	Christie's Chicago	Butterfield's San Francisco	Total
Later price higher	271	143	90	20	524 (11%)
Later price lower	628	430	183	41	1282 (28%)
Later price equal	1498	1073	226	39	2836 (61%)

	Chateau Palmer 1961			Croft 1927			Chateau Margaux 1952			Quinta de Nora 1934		
	Lot size	Price	Price/bottle	Lot size	Price	Price/bottle	Lot size	Price	Price/bottle	Lot size	Price	Price/bottle
Lot 1	12	920	77	12	800	67	12	480	40	10	400	40
Lot 2	12	800	67	12	800	67	12	480	40	12	500	42
Lot 3	12	700	58	12	750	63	12	480	40	12	500	42
Lot 4				12	650	54	24	480	20	12	480	40
Lot 5				12	650	54	24	480	20	12	480	40
Lot 6				12	650	54	20	480	24			
Lot 7				12	650	54						

Source: Liquid assets, *International Guide to Fine Wines* (issue 4, spring), 1988.

these auctions as "the repeal of the law of one price." This phenomenon has been the subject of numerous theoretical and empirical studies.[3]

This lack of consistency in prices was noted in a study of wine auctions and is illustrated in table 6.1.[4] The first auctions for the 1961 Palmer and 1927 Croft each sold for a significant premium with respect to the final auctions. These data appear inconsistent with standard models of perfect competition. Also other empirical evidence suggests what is often called an *afternoon effect* or a *declining-price anomaly*; that is, as in these wine auctions, for the early lots to sell at a premium relative to the later lots is fairly common, and much more common than for prices to increase across auctions.[5] The task addressed in what follows is to identify the factors that determine what patterns of prices arise in practice.

In theory, there are two effects. One is that bidders should want to bid less aggressively for the first lots than for the later lots. In the first auction there are more remaining chances to win. In the last auction, bidders should be willing to bid up to their reservation values, but not in the other auctions. This logic means that a bidder's optimal bidding strategy should be to bid an increasing fraction of its reservation price from one auction to the next. However, the countervailing factor is that the first winners will tend to be those with the highest valuations, so competition should tend to decrease from one auction to the next. In theory it is possible for these two effects to cancel out. One main result of this chapter is that under certain conditions, they do. This chapter also explains some conditions under which price can be expected to increase or decrease.

This chapter also explains when prices can be expected to increase or decrease across auctions. Briefly, when a seller (a buyer) holds two equivalent auctions, then the buyers (sellers) bidding in the first auction will want to offer a slightly lower (higher) price than the expected value of the price in the second auction. The intuition behind this result is that a bidder in the first auction will want to pay no more than the expected price in the second auction less an amount to adjust for the expected surplus from winning the second.

6.3 Weber's Martingale Theorem

This section presents a fundamental and surprising result, Weber's *martingale theorem* characterizing the expected values of prices in a sequence of auctions. This result states that the expected price in one auction is, under conditions explained below, equal to the price in the previous auction. In other words, the most recent auction price is the best predictor of the price in the next auction. The intuition behind this result derives from the fact that if strategic bidders were to think one auction would produce a lower price, these bidders would all want to participate, and remain active, in that auction until the price reached the level of other auctions. This strategic behavior would tend to arbitrage the prices across auctions. In this case the auction design, and more specifically the allocation of the total quantity across auctions, should not affect the expected price in any auction or the overall auction revenues.

Consider a model in which there are some number B of bidders, and $I < B$ items for sale. Each bidder can win at most one lot. To simplify the exposition, it is assumed that each bidder has a valuation v_b and that the bidders' valuations are uniformly drawn from a uniform distribution over the unit interval. Assume that bidders are labeled so that $v^b > v^{b+1}$ for all $b = 1, 2, \ldots, B - 1$. These assumptions allow an easy characterization of equilibrium, and as will be clear from what follows, the results apply quite generally. What follows assumes that the lots are sold in a sequence of first-price auctions.

The following basic insights allow a direct calculation of the solution to the auction:

1. *Backward induction* To solve for the equilibrium of this auction, it is easiest to start working backward, from the last auction. First, assume that all but one item has been sold, and that there are $B - I + 1$ bidders remaining.
2. *Revenue equivalence theorem (RET)* A consequence of the RET is that the highest value remaining bidder, I, should win the last auction and have to pay a price equal to the value of the second highest remaining bidder. So the price that the high-value remaining bidder pays should equal v^{I+1}, the value of the $I + 1$st remaining bidder.
3. *Order statistics* The expected value of v^b, the bth highest valuation or the bth order statistic, is $[(B+1-b)/(B+1)]$. This means that the expected value of v^{I+1} is $(B - I)/(B + 1)$. Notice that this ratio is close to 1 if I and B are large and that I is almost as large as B. If $I = \alpha B$, $\alpha \in (0, 1)$, and both I and B are large, then the price will be approximately $1 - \alpha$.

The evaluation above implies that last item sold will sell for a price of $(B - I)/(B + 1)$: assuming that this last auction is a second-price sealed-bid auction. (Note that the RET implies that the item would also sell for this same price if the last auction were a first-price sealed-bid or an oral ascending auction.) Bidder I's value will, on average, be $(B - I + 1)/(B + 1)$: and I will pay $(B - I)/(B + 1)$:, which is the fraction $(B - I)/(B - I + 1)$: of I's value. Then the question is what should be the price of the next to last item sold. Assuming that the $I - 2$ highest value bidders win items 1, 2, …, $I - 2$, then the remaining bidders will be $I - 1, I, I + 1, \ldots, B$. Bidder I should win that last auction but not pay more than $(B - I)/(B + 1)$: the same price as bidder I. Therefore, in the next last to auction, bidder $I - 1$ should win on offering a price that equals the fraction $(B - I)/(B - I + 2)$: of its value.

By backward induction, all I objects sell for the same price$(B - I)/(B + 1)$: and the bidders should offer increasing fractions of their value, $(B - I)/B, (B - I)/(B + 1), \ldots, (B - I)/(B - I + 2), (B - I)/(B - I + 1)$. In this example, the expected price is constant, and in particular, the expected value of the price in each auction is the same as the price in the last auction. This property means that the price is what is called a *martingale*. This result is quite a bit more general.[6]

Proposition 1 *Suppose that there are I identical items available in a sequence of first-price auctions. Suppose that each bidder can purchase one item, and that a bidder with valuation v will bid more for an item in an auction than a bidder with valuation $w < v$. Then the expected value of the price in each auction will equal the price in the previous auction, that is,* $E(p_i \mid p_{i-1}, p_{i-2}, \ldots, p_1) = p_{i-1}$.

Sketch of Proof Let I_j denote the information available when item $j + 1$ is to be sold. Notice that RET implies that the expected payment by winners of the last $I - j$ lots is $E[v^{I+1} \mid I_{j+1}]$. Then the expected price as of auction j for items sold in subsequent auctions $j + 1, j + 2, \ldots, I$ is $E[E[v^{I+1}] \mid I_{j+1} \mid I_J] = E[v^{I+1} \mid I_j]$. Therefore this price must be the same as the price for auction j. ■

Notice that the martingale theorem extends to the case where the auctioneer can divide the available volume across any fixed number of auctions. More specifically, suppose that the auctioneer conducts a sequence of uniform-price auctions. In each auction a, l_a lots are sold. Also suppose that in each auction the top l_a bidders win and pay the lowest of the winning offers. Then a similar argument implies that all items will sell for a price equal to v^{I+1}, independently of the how the items are distributed across lots.

6.4 Strategic Allocation Decisions

In practice, auctions can include multiple units sold (or purchased) at one time. Thus an auction originator can face the decision how to divide the amount available across auctions. The martingale theorem suggests that decisions about how to allocate quantity sold (or purchased in a reverse auction) across auctions should not affect the expected outcome.

There are a variety of reasons why this is not necessarily the case. One of the most common is uncertainty about overall market conditions. For example, in a series of ongoing energy procurement auctions, the utilities conducting the auction leave open the option of limiting procurement volume if transitory market conditions are highly unfavorable. Transitory energy price spikes that occur on the day of an auction could leave the utilities, and the ratepayers who are their customers, paying prices that reflect the price spikes, not the underlying market conditions. Unusual events, such as hurricanes, can create significant short-term price uncertainty, which can justify some auction volume reductions.[7]

This section explains two other factors. One is how the split of the auction volume across time can affect relative competition, bid strategy, and prices across auctions, and the other is how resolution of uncertainty may affect the outcome.

These questions are related to the question of the strategy a buyer or seller should take when choosing between forward and spot market transactions. Allaz and Vila (1993) consider a model in which firms can compete in both spot and forward markets. The offers in the forward market affect the payoffs in the spot market. This can be modeled as a two-stage game, in which the firm choose quantities at each stage and —the first-stage sales are not made to the highest value buyers but to a subset of *average* buyers over the two stages. What Allaz and Vila find is that if one firm participates in the forward market and the other cannot, then the one that participates will be able to earn a higher profit.[8] If both try to compete in a forward market, then both earn less than if they had waited for the spot market. As is explained below, Allaz and Vila's (1993) result has a natural extension to a sequence of auctions.

6.4.1 A Simple Two-Auction Example

This subsection considers a very simple example in which the auctioneer wants to sell two objects one at a time to B bidders.[9] Bidder valuations *in each auction* are uniformly distributed on [0, 1]; that is, the distributions of the bidder valuations in the two auctions are stochastically equivalent. This means that each bidder's valuation for each object in each of the two auctions has the same distribution. This does not mean that any bidder's valuation in one auction is the same as its valuation in any subsequent auction. So the highest value losing bidder in the first auction will not necessarily be the highest value bidder in the second auction. This is in contrast to the assumption of the martingale theorem, that bidder valuations do not change from one auction to the next.

More specifically, it is assumed that each bidder obtains a signal of its value for each object before each auction starts and that the distribution for determining values is the same across auctions (and bidders). This type of model allows for variations where, for example, the lots vary slightly in ways that each bidder might value in a different way, or where bidders' circumstances can shift in a manner which is equally likely to cause value to increase or decrease. In what follows, a few simple examples are used to illustrate how shifts in bidder valuations across auctions can affect whether prices tend to increase or decrease.

A Two-Unit Example As a base case example, first recall the example above in which there are B bidders competing for two (rather than I) identical objects, and that bidder values do not change at all across the two auctions. It is assumed that each bidder is interested in only one object. For the sake of illustration it will be assumed that bidder valuations are fixed before any of the auctions start and are drawn from a uniform distribution on [0, 1]. On one hand, if the objects are sold in a single first-price (i.e., pay-as-bid) auction, then the bidders with the highest two values will win, and the price will be $(B-2)/(B+1)$, which is the expected value of the third-highest value bidder.[10] On the other hand, if the two objects are auctioned one at a time, then there would be $B - 1$ bidders competing in the second auction. Their expected values would be

$$\frac{B-1}{B+1}, \frac{B-2}{B+1}, \ldots, \frac{1}{B+1}.$$

The expected price would be

$$\frac{B-2}{B+1}.$$

Therefore in the first auction no bidder would want to bid above

$$\frac{B-2}{B+1},$$

and so this would be the expected price in that auction as well. Thus expected prices are the same whether there is one auction for two objects or two auctions for one object each.

Now suppose that bidders' values are not necessarily the same for the two objects. This could be the case, for example, if bidders acquire information or their opportunities changes between the two auctions. The latter will occur when, for instance, bidders may engage in other related transactions. Or it can be the case that bidders view the two items as very similar, and bidder valuations for each item are the same except for some random component that is not correlated across items for each bidder. That is, bidder valuations are *stochastically equivalent*.

As above, the auctioneer has two items for sale, and can choose to sell them either all at once or sequentially. The B bidders each can still win at most one item, but now each bidder's valuation for an item is a separate random draw in each auction from the uniform distribution over the unit interval. Bidders also only learn their values shortly before each auction starts.

In this case it is easy to compute the expected value of the price if both items are sold at one time in an auction in which both winning bidders pay the third highest offer amount[11]:

$$p^2 = \frac{B-2}{B+1}.$$

Then expected auction revenues will be

$$R^2 = \frac{2B-4}{B+1}.$$

If, however, the items are sold one at a time in two second-price auctions, then the expected price in the second (last) auction will be

$$p_2^{1+1} = \frac{B-2}{B}.$$

This means that the expected surplus derived from participating in the second auction will be

$$\frac{1}{B(B-1)}.$$

Therefore (1) a bidder in the first auction will not want to bid up to its full value, as it has a positive expected value from participating in the second auction, and (2) the highest value loser in the first auction cannot always expect to win the second unit. Thus the expected value of the price in the first auction will be

$$p_2^{1+1} = \frac{B-1}{B+1} - \frac{1}{B(B-1)}.$$

So expected revenues from selling the items one at a time will be

$$R^{1+1} = \frac{B-1}{B+1} + \frac{1}{B-2} - \frac{1}{(B-1)B}.$$

Now

$$R^{1+1} - R^2 > \frac{1}{B(B+1)}(B-3) > 0$$

for $B > 3$. Therefore selling the items sequentially raises more revenue when there are more than three bidders.

A Three-Unit, Two-Auction Example A variant of the second example considers the case with three units and two auction dates. More specifically, suppose the auction originator wants to sell three units in two auctions. It has the option of selling a fraction x in the first and y in the second. In this case the expected prices in the two auctions depend on whether one unit or two is sold in the first auction.

If one unit is sold in the first auction, then the expected valuations in the second auction of the remaining $B-1$ bidders are $\frac{B-1}{B}, \frac{B-2}{B}, \ldots, \frac{1}{B}$. This means that the expected price in the second auction will be

$$E(p_2^{12}) = \frac{B-3}{B}.$$

Thus those bidders losing the first auction can anticipate a surplus of $1.5/B$ with a probability of $2/(B-1)$. So, in the first auction, bidders will shade their bid by

$$s^{12} \equiv \frac{3}{B(B-1)}.$$

Hence the expected price in the first auction will be

$$E(p_1^{12}) = \frac{B-1}{B+1} - \frac{3}{(B-1)B}.$$

Similarly, if two units are sold in the first auction, then the expected valuations in the second auction of the remaining $B-2$ bidders are $\frac{B-2}{B-1}, \frac{B-3}{B-1}, \ldots, \frac{1}{B-1}$. The expected price in the second auction will be

$$E(p_2^{21}) = \frac{B-3}{B-1}.$$

This means that the expected surplus is

$$s^{21} \equiv \frac{1}{(B-1)(B-2)}.$$

So in the first auction the expected price will be

$$E(p_1^{21}) = \frac{B-2}{B+1} - \frac{1}{(B-1)(B-2)}.$$

Comparing costs $R^{21} \equiv E(p_1^{21}) + 2E(p_2^{21}) - 2s^{21}$ with $R^{12} \equiv E(p_1^{12}) + 2E(p_2^{12}) - s^{12}$ indicates that $R^{12} - R^{21} = [2/(B+1)][1-(1/B)] + s^{12} - 2s^{21}$. This is necessarily positive for large B.

The two examples above illustrate three points:

- First, the fact that bidder valuations can vary from auction to auction, although they have the same distributions, means that there is an option value from participating in the last auction that does not exist when bidder valuations are constant across auctions.
- Second, the degree of competition will vary across auctions. The first-order effect means that the larger the number of remaining bidders relative to the remaining auction volume in the last auction, the lower is the price in the last auction.
- Third, when the auction originator has some flexibility about how to divide the total volume available, then decisions about how the volume is divided, and (as will be explained below) the flexibility of the auction originator to shift volume across auctions can affect average auction prices and the combined proceeds from the auctions.

This logic generalizes to some extent to multiple auctions and multiple lots. When bidder valuations across auctions vary, but are stochastically equivalent, the allocation of the auction volume across auctions will affect the expected price in the final auction, and therefore earlier auctions. The relevant order statistic for determining the last auction price is $(B-I+x-1)/(B-I)$, where I is the number of items, B the number of bidders, and x the number of lots sold before the final auction. This price will depend on x.

6.4.2 Optimizing the Distribution of Auction Target Volume across Auctions

This section considers a generalization of the preceding example with two auctions. It considers a reverse auction. In it the auctioneer wants to *purchase* a total of I units from N bidders, and can choose how to divide this into X units in the first auction and $Y = I - X$ in the second. The question addressed here is what is the optimal way to do so to minimize expected overall procurement costs. The analysis of how to divide a total number of lots to be sold across two auctions is essentially the same.

Given that there are two auctions, and bidders' costs in each auction are, as above, a random, independent, and uniform draw over the unit interval, the second-auction price will be

$$p_2 = \frac{Y+1}{N-X+1} = \frac{I-X+1}{N-X+1}.$$

This implies that the average second-period winner will have costs of

$$\frac{Y+1}{2(N-X+1)},$$

and will therefore have an average surplus (for winners) of

$$\frac{Y+1}{2(N-X+1)}.$$

The first-period price will then be

$$\frac{X+1}{N+1} + \frac{Y}{(N-X)}\frac{Y+1}{2(N-X+1)} = \frac{X+1}{N+1} + \frac{(I-X)}{(N-X)}\frac{I-X+1}{2(N-X+1)},$$

the probability that a first-auction loser will be a second-auction winner is

$$\frac{I-X}{N-X}.$$

Total costs are then

$$X\left(\frac{X+1}{N+1} + \frac{I-X+1}{2(N-X+1)}\frac{(I-X)}{(N-X)}\right) + (I-X)\left(\frac{I-X+1}{N-X+1}\right).$$

Notice that the first-auction expected price is increasing in the quantity purchased in the first period, and similarly that the second-auction price is increasing in the quantity purchased in the second auction. The optimal ex ante value of $X \in \{0, 1, \ldots, I\}$ will minimize

$$C(X, I) = X\left(\frac{X+1}{N+1} + \frac{I-X+1}{2(N-X)(N-X+1)}\right) + (I-X)\left(\frac{I-X+1}{N-X+1}\right), \tag{6.1}$$

and, if there is no second auction, then the second term in the right-hand side of (6.1) is zero. Therefore

Table 6.2
Sequential auctions of 4 lots to 10 bidders

x	y	p_1	p_2	Total costs
0	4	—	0.46	1.82
1	3	0.25	0.4	1.45
2	2	0.31	0.33	1.29
3	1	0.38	0.25	1.39
4	0	0.46	—	1.82

Table 6.3
Sequential auctions of 250 lots to 400 bidders

X	y	p_1	p_2	Total costs
0	250	0.198	0.63	156.48
100	150	0.377	0.501	112.98
140	110	0.442	0.425	108.60
141	109	0.443	0.423	108.598202
142	108	0.445	0.421	108.598178
143	107	0.446	0.419	108.60
145	105	0.449	0.414	108.63
150	100	0.457	0.402	109.79

$$C(0, I) = C(I, I) = I\left(\frac{I+1}{N+1}\right).$$

It will be optimal, ex ante, to divide the auction quantity between the two dates.

Tables 6.2 and 6.3 provide some illustrative calculations. The first example assumes $N = 10$ and $I = 4$; the second example considers the case where $N = 400$ and $I = 250$.

These examples show that there is a benefit in splitting the auction quantity over the two auctions, and some advantage in auctioning more in the first auction than in the second, at least when there is a large quantity to be purchased in aggregate. The intuitive reason to divide the auction quantity is that the auction manager gets a larger sample of offers. A bidder at one date is effectively a different bidder at a different date, in that its costs can be higher or lower. Dividing the volume lowers the expected procurement costs, absent strategic bidding. However, bidders will behave strategically. Bidders in one

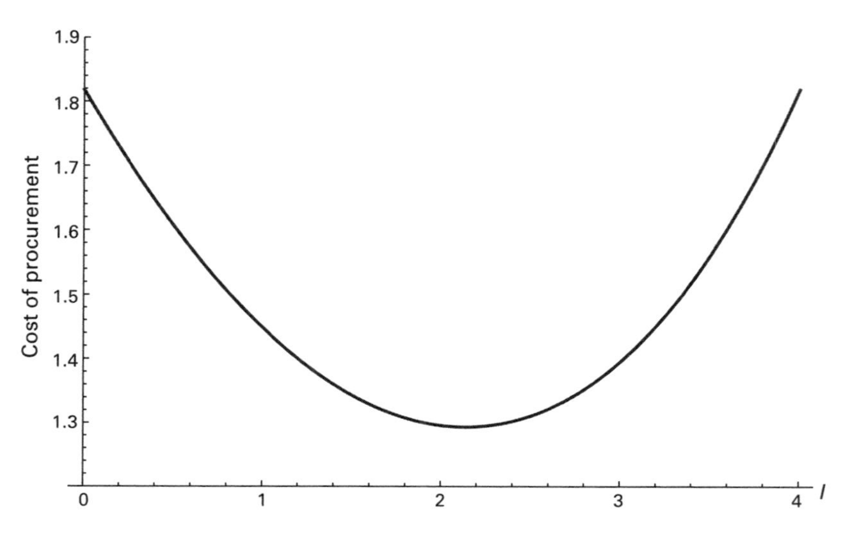

Figure 6.1
Ten bidders and 4 lots

auction will want a higher profit margin than in a later auction. This strategic withholding effect limits how much the auction manager will want to delay procurement.

These examples also assume that the decision to split the auction volume is made before the first auction, and before any bids are received. If the auctioneer can defer the decision to determine the auction volume on each date based on the number of bidders, then it will be possible for the auctioneer to obtain a still better outcome. More specifically, given any volume that the auctioneer might choose to leave for the second auction, it can calculate the expected auction price. This price is what it should use to set reservation prices in the first auction. Such reservation prices in the first auction can only improve its outcome.

6.4.3 Sequential Auctions and Risk

A bidder's preference toward risk can be a factor that enters into a decision about how much to bid in one auction when there are other auctions to follow.[12] Consider the case where, in a forward auction, a bidder can make a purchase in one of two auctions, each of which is a second-price auction. If the high-value bidder in the first auction lets the second high-value bidder win by bidding low, then in the second auction it can expect to win the object and pay the third highest value. Let $x - V^3$ be the money the bidder spends on other goods after paying

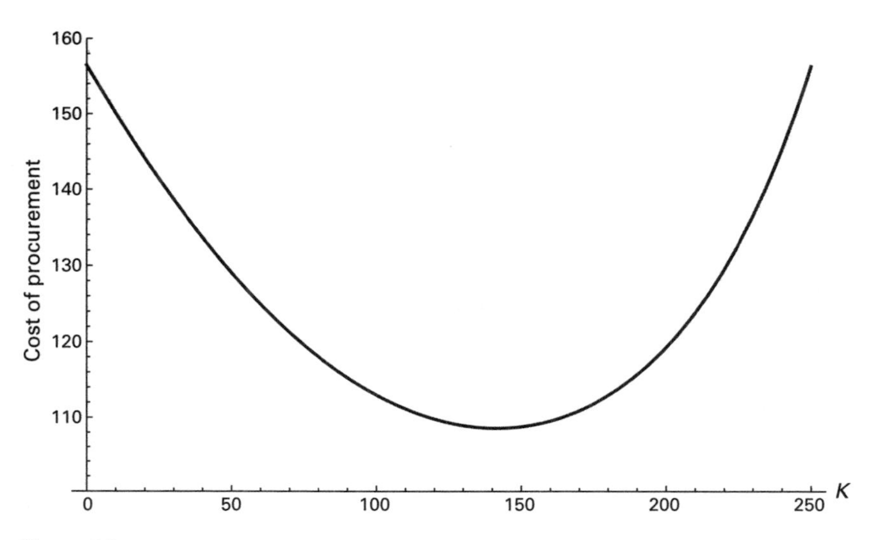

Figure 6.2
Four hundred bidders and 250 lots

V^3, where V^3 is the (random) value of the third highest value bidder. This will provide the bidder, ex post, with some utility $u(x - V^3)$ in addition to the value of the item.[13]

Suppose that this bidder is risk averse, and could win the object for the expected value of $V^3 = v^3$. Because of risk aversion, $u(x - v^3) > E[u(x - V^3)]$. As this is the case, the bidder would offer more in the first auction than v^3 to ensure against the risk associated with the randomness of the second auction price. Thus, if all bidders are risk averse, then they should each offer a bit more than v^3 to win the first auction, and the expected price should be falling.

If risk aversion is the determining factor in price dynamics across auctions, then prices should decrease. If changing bidder opportunities are then prices can increase or decrease. Thus, when there are multiple auctions over time, then prices sometimes increasing and sometimes decreasing suggest that risk aversion alone is an inadequate explanation of price patterns. However, shifts in bidder valuations across auctions would suggest that the price patterns would depend on how the total volume is divided across auctions.

6.4.4 Auction Timing

This subsection now provides another example of how auction timing can affect the auction revenue when bidders learn more about values over time. In particular, it is assumed that there are two bidders and one item for sale. The value each bidder will place on the item is either 0 or 1, and the two bidders have a common ex ante prior, $\alpha \in [0, 1]$, that each value will be high. Moreover the valuations are independently and identically distributed. So there are four possible outcomes (0, 0), (1, 0), (0, 1), and (1, 1) with probabilities of $(1 - \alpha)^2$, $\alpha(1 - \alpha)$, $\alpha(1 - \alpha)$, and α^2, respectively. Suppose that the item can be auctioned at one of two dates, early or late.

An early auction means that no ex ante uncertainty is resolved. If the item is sold at the early date, then the price, in either a first-price or a second-price auction, will be α. The winning bidder has zero expected surplus. If, in contrast, the auction is conducted at the late date, then the price will be 1 with probability α^2. In a second-price auction—or, by revenue equivalence, in a first-price auction—the late auction price will be 0 with probability $1 - \alpha^2$. Therefore the expected revenues are α^2. Notice that the late auction expected revenues α^2 are less than α, the expected revenues in the early auction.

This result is reversed with a larger number of bidders. If there are B bidders, then the probability that no more than one bidder will have a value of 1 is $(1 - \alpha)^B + B\alpha(1 - \alpha)^{B-1}$, and the expected value in a late auction is $1 - [(1 - \alpha)^B + B\alpha(1 - \alpha)^{B-1}]$. The late auction will have a higher or lower expected value depending on B and α. When $\alpha = 0.5$ and $B = 3$, then an early and a late auction have the same expected revenues.

6.5 A Summary of Auction Timing Considerations

This chapter considered situations in which different units of an identical product can be purchased or sold at different dates. This is a fairly common occurrence in many industries, ranging from wine to electricity. The way in which the different units are divided over different auctions at different times can, but need not, affect the outcome. Indeed a very fundamental result is that strategic arbitrage by bidders will tend to equalize prices across auctions. When this *martingale* result holds, auction design and bidding strategy are both simplified. However, the martingale theorem will not always apply.

When the martingale theorem does not apply, the possibility of shifting purchases and sales over time confronts both bidders and the auction originator with strategic choices. When bidder valuations tend to shift, the examples above suggest there is a tendency of prices to decline in forward auctions, or increase in reverse auctions. However, the auction originator can choose to allocate quantities in a way that can cause prices to either increase or decrease. And competition can shift over time. Therefore bidders cannot plan on prices falling.

The results above suggest that an auction originator's choice of timing will affect the outcome, even without uncertainty. When there is uncertainty, the auction manager will also want to consider the impact of the resolution of uncertainty on competition and the expected prices.

Thus a firm, government body, or other entity considering auctioning off a set of objects should consider a number of factors in determining whether and how to auction the objects in different events. The relative benefits of different options will depend, in part, on how the allocation affects bidder strategies, as well on how it affects the intensity of competition across auctions. As can be seen above, these choices can affect the outcome in favor of the auctioneer or in favor of the bidders.

7 Sequential Auctions of Complements

Summary

This chapter examines auction design and bidding decisions in sequences of auctions in which the outcome of one auction affects a bidder's profits from winning a subsequent auction. At times, a winner of an initial contract or concession will have a stronger incentive to extend its contract or renew its concession in a second auction than it would had it lost the initial auction. This chapter also explains how the bidding strategy in one auction can be affected by the effect the outcome of that auction may have on a follow-up auction for a complementary product or contract. In both cases we find a tendency toward increasing dominance; that is, the winners of an initial auction have an advantage in follow-up auctions. This also causes bidding to be more competitive in an early auction and less competitive in a subsequent auction.

The complementarities considered can be of two forms. First, the winners of one lot may have a higher value, or a lower cost, for a complementary lot. Second, a winner in one auction may, possibly over time, invest in complementary assets that give it advantages in subsequent auctions.

7.1 Examples of Complementarities

Complementarities are common in auctions. They arise out of decisions of the auction designer about how to divide what is to be purchased or sold into lots. There are many reasons why an auction manager may not want to have one large lot. One primary reason is that different bidders will have different interests. Bidders may be seeking overlapping sets of lots. Or there may be small bidders and large ones. For this reason an auction may include both large and small lots. This chapter assumes that the lots are sold in sequence; for the most part it also assumes that the lots are uniform in size.

There are plenty of examples of sequential auctions with complementarities across regions. The European 3G and 4G spectrum auctions were sequential. At least for some of the bidders, the value of a spectrum license in one EU member country would be positively correlated with having licenses in other, nearby EU member countries. There are numerous reasons for this, including the fact that knowledge about the technology and network equipment can be transferred across borders, economies of scale in obtaining in marketing, roaming advantages, and economies of scale in network development and management. In addition most the 4G auctions included blocks in multiple spectrum bands, which could be either substitutes or complements.

The other type of synergy of concern here is value dependences over time. One fundamental question in many auctions for concessions, licenses, or procurement is the duration of the contract. In many cases, to derive value (in a forward auction) or meet performance requirements (in a reverse auction), the winners must make sunk investments in durable capital projects whose economic life far exceeds the term of the contract, license, or concession. It is often argued that short-duration contracts will not allow the bidder to recover fixed investment costs, or that bidders will need to inflate costs to be able to recover their fixed investments. However, this assertion can only be valid if the winners in an initial auction do not win subsequent renewal auctions. As we will see, it is often the case that initial winners will win follow-up auctions. Therefore it is far from obvious what effect the initial contract duration will have on the outcome. Renewal auctions are quite common in some sectors.

One example of this type of problem is spectrum auctions; most spectrum licenses are for a limited term, ranging from ten to twenty years.[1] The question arises as to what happens after the initial license duration or contract ends. While in some cases, as in the FCC auctions, there is a *presumption of renewal* of the contracts, this is far from universally true, and the terms of renewal are often not even described, much less fully specified. It can also take five years or longer to fully build out a license area, and once it is built out, the facilities, such as radio towers, cables connecting towers to the network, and other network equipment, will have value beyond the license term. Moreover a firm will not want to turn off service to its customers the day its license expires, but it could, in theory, be faced with this prospect.

In the energy sector, in many geographic areas, regional transmission organizations (RTOs) have been organizing bidding procedures for the procurement of capacity. In a capacity auction the bids are for payments beyond anticipated energy generation service revenue, to ensure capacity gets constructed that otherwise might not be.[2] The bids often cover a period of no more than ten years, and usually less, despite the fact that the facilities often have useful lives of twenty or more years. These auctions are being conducted both for resource adequacy (i.e., to ensure adequate reserve margins) and for renewable-energy procurement. Similar auctions have been planned and conducted for transmission capacity. If the contracts are for ten years but the facility lasts twenty years, then the bidder may have an incentive to ask for nearly twice as much, per unit or per year, for a ten-year contract as it would for a twenty-year one. This logic fails if the first winner is almost certain to win an auction for a contract renewal. However, a shorter contract term does create some risk of losing the contract part way through recovering up-front investments, which is a risk that does not arise with longer contracts.

One auction design option when there are complementarities of the types described above is to permit packages—either prespecified, or allowed to be specified by the bidders. This is often not a practical option. One reason is that package bidding introduces added complexities discussed in chapter 10. Second, different jurisdictions may not be able to, or may not want to, conduct a simultaneous auction or to create packages. For example, it might be impossible for EU member countries to agree on a single auction for all their 3G or 4G licenses. Third, when the complementarity is over time, the early winners may have strategically different positions than rivals in renewal auctions. The remainder of this chapter considers complementarities in sequential auctions.

7.2 Renewal Auctions

This section examines the influence of contract duration in a sequence of auctions. The section starts with a simple example. It is assumed that there is a franchise auction at some date 0, and a follow-up auction at some other date 1. Each auction is for a fixed-duration contract—for example, a spectrum license for a fixed number of years. It is assumed that any winner in the first auction must make a sunk durable investment of $K > 0$ to extract value, but that this investment will not depreciate at all through the end of the second contract period. With the investment of K, a bidder $j = 1, 2, \ldots, N$ can expect to earn $v_j - K$ in the first period if it wins the first auction, or in the second period if it has not already incurred a first-period investment of K; and to earn v_j in the second period if it (won the first auction and) has already made the sunk investment of K. In what follows, it is assumed that $2v_1 > K$; that is, there is a positive surplus available.

Further suppose that there are the same $N \geq 2$ bidders in each auction. To simplify the exposition, it is assumed that each auction is a second-price sealed-bid auction. Solving for equilibrium is straightforward.

Start at the last auction. Suppose that bidders are indexed, so that $v_j \geq v_{j+1}$ for all $j = 1, 2, \ldots, N - 1$. Also suppose that $v_1 - v_N < K$; that is, the difference in bidder values is not as large as the sunk costs. Then, if bidder 1 wins the first auction, it will be willing to offer v_1 in the second auction, and the next best offer would be from bidder 2, which would be willing to offer $v_2 - K$. By assumption, $v_2 \leq v_1$, so $v_1 > v_2 - K$, and bidder 1 will win the first auction, at a price of $v_2 - K$.

If some other bidder, say bidder h, has won the first auction, then $v_1 - v_h < v_1 - v_N < K$, by assumption, so that bidder h will win the second auction, at a price of $v_1 - K$. Thus the bidder that wins the first auction will win the second auction.[3] And the second auction price will be $v_1 - K$ if bidder 1 did not win the first auction, or $v_2 - K$ if bidder 1 did win the first auction.

This allows solving for the price, and winner, in the first auction. Bidder 1 would pay up to

$$b_{11} = v_1 - K + [v_1 - (v_2 - K)] = 2v_1 - v_2 \tag{7.1}$$

in the first auction, and the most any other bidder would offer in the first auction, which would be offered by bidder 2, is[4]

$$b_{21} = v_2 - K + [v_2 - (v_1 - K)] = 2v_2 - v_1, \tag{7.2}$$

where b_{ij} is the amount bidder i offers in auction j. Notice that the term $[v_1 - (v_2 - K)]$ in (7.1) is bidder 1's expected surplus in the second auction. As $v_1 > v_2$ by assumption, $b_{11} > b_{21} = 2v_2 - v_1$. This is an equilibrium outcome for the first auction provided that $2v_2 - v_1 \geq 0$. If $2v_1 - v_2 > 0 > 2v_2 - v_1$, then bidder 1 would win the first auction and have to pay $\max\{0, 2v_2 - K\}$, which is the most that bidder 2 could possibly offer. The amount bidder 1 pays in the first auction, $2v_2 - v_1$, exceeds bidder 1's profits in the first period, $v_1 - K$ whenever $K > 2(v_1 - v_2)$.

Bidder 1 is willing to lose money at first because it can recover the losses in the second period. Notice too that the price in the second auction is higher (lower) than in the first when $v_2 < (>) v_1 - K$.

This logic extends to the case where there is a sequence of S auctions, each for a fixed contract duration. In what follows, it is assumed that the fixed and sunk investment is the same as above, and will have an economic life at least as long as that which is spanned by the S auctions. Solving for equilibrium can be accomplished in a manner similar to the above, starting with the last auction.

There are several possibilities in the last auction. Suppose, without loss of generality, that firm 1 is the firm with the highest value of v_j that has won a previous auction. Then firm 1 will have a higher valuation for the last auction, $v_1 > v_j > v_j - K$ for all $j \neq 1$. And the most some other firm will pay is either v_j or $v_j - K$ for some $j \neq 1$. So the highest value firm that is a previous winner going into auction S will win auction S and pay $\max_{j \neq 1}\{v_j\}$ over previous winners j, or $\max_j\{v_j - K\}$ if firm 1 won all previous auctions.

Working backward to auction $S - 1$, we find that the highest value winner of the previous auction $S - 2$ should win auction $S - 1$ (unless K is not very large and the difference between v_1 and the j that was the highest value previous winner is relatively large). The willingness to pay of the highest value winner in auction S is $v_1 + (v_1 - (v_2 - K)) = 2v_1 - v_2 + K$, assuming 1 won all previous auctions, or $v_1 + (v_1 - v_h) = 2v_1 - v_h$, assuming h is the highest value previous winner other than firm 1. Both are larger than $2v_h - v_1$, which is the most another bidder would offer in auction $S - 1$. Inductively, the highest value winner in auction $S - t$ will win auction $S - t - 1$, and so all auctions will be won by the same bidder, bidder 1. Bidder 1 will derive a surplus of at least $v_1 - v_2$ in each of auctions 2, 3, …, S, and $v_1 - v_2 + K$ along the equilibrium path. So, in auction 1, bidder 1 will be willing to offer at least $v_1 + (S - 1)(v_1 - v_2) - K$, and should be willing to offer as much as $v_1 + (S - 1)(v_1 - \max\{0, v_2 - K\})$. The next-highest value bidder, 2, should similarly be willing to offer as much as $v_2 + (S - 1)(v_2 - \max\{0, v_1 - K\})$. So the first auction can be for a very high price, larger than $\min\{(S - 1)K, (S - 1)v_2\}$, with the losses in the first auction offset by profits in the renewal auctions.

Notice that in a sequence of auctions in which the first auction's winner has an advantage in subsequent auctions, the competition in the first auction can be intense. Prices can far exceed the franchise value. However, the losses in the initial contract are offset by the profits from the subsequent renewal contracts. Notice too that the contract duration does not affect overall auction revenues. It does affect auction prices—longer contracts reducing the first-auction price, and not increasing it.

7.3 Sequential Auctions of Complements

This section considers a sequence of auctions for complements. More specifically, it is assumed that the sum of bidder stand-alone valuations for individual objects is less than the valuations for the packages. Arguably, this is often the case in spectrum auctions, in which bidders realize economies of scale and scope when acquiring more spectrum in a larger geographic area. For example, a firm may place a higher value on a UK license if it also wins licenses in Germany and Italy. Or a firm may place a higher value on a New York license if it also wins a Chicago and a Los Angeles license. There is a large literature on auction design for complements and for package bidding. However, oftentimes it is not possible to conduct auctions for complements at one time or to allow package bidding.

A very simple example can illustrate how complementarities can affect bidding decisions and prices. Suppose that there are two bidders and three lots, and the lots are sold in a sequence of first-price auctions. Further suppose that each bidder derives a value of 1 from two or three lots, and 0 from one lot. The equilibrium price in the first auction of this "chopsticks" example will be 1, and the prices in the second and third auctions will be 0.

Or consider the following straightforward extension of the example from the previous section. Suppose that there are two lots, $k = 1, 2$, and bidders $j = 1, 2, \ldots, n$. Suppose too that for each bidder j, the value of winning lot k is v_{jk} and the value of winning both is $v^j > v_{j1} + v_{j2}$. That is, each bidder places a higher value on the package than the sum of the individual lot values. In what follows it is also assumed that $v^j > v^{j+1}$ and $v_{jk} > v_{j+1k}$ for all j and k.

Consider a sequence of two second-price auctions. To solve for equilibrium, start with the second auction. There the first auction's winner will have a value of $v^j - v_{j1} > v_{j2}$, and other bidders i have a value of v_{i2}. If bidder 1 wins the first auction, then it will win the second, as by assumption $v^1 - v_{11} > v_{12} > v_{i2}$ for $i > 1$. If bidder 1 does not win the first auction, the first auction's winner j will win unless $v_{12} > v^j - v_{j1}$.

So, if 1 wins the first auction, it can expect to win the second auction at a price of v_{22} and derive a surplus of $v^1 - v_{22}$. Therefore 1 should be willing to bid up to this amount to win the first auction. No other bidder j can derive a greater surplus in the first auction. The most another bidder can expect to win is $v^2 - v_{12}$, as by assumption bidder 2 has the next highest valuations and would at most earn a surplus of $v^2 - v_{12}$ from winning the first auction. Thus it is a subgame perfect equilibrium for bidder 1 to win both auctions, the first at a price of $v^2 - v_{12}$ and the second at a price of v_{22}. Note that bidder 1' s value minus the total amount paid by bidder 1, $v^1 - v^2 + v_{12} - v_{22}$, is positive, as by assumption $v^1 > v^2$ and $v_{12} > v_{22}$.

The argument above will extend to a larger set of complements. More generally, the first auction's winner will tend to have an advantage in auctions for complements. In the example above, the first-auction price can be quite high relative to the second-auction price. For example, suppose $v^k = (N - k+2)v = (N - k + 2)v_{k2}$ for $k = 1,2$ in the example above; then the price in the first auction will $(N - 1)v$, and the price in the second auction will be v. Thus for strong complements, namely large N, the price in the first auction will be many times the price in the second.

In the European 3G auctions there were some complementarities—both on the cost side, in rolling out service, and on the revenue side, at least from roaming revenues. Firms with operations in one country can use the same equipment vendors in another country, and usually benefit from economies of scale. Also European mobile operators are able to capture a higher share of high-margin international roaming when they have operations in multiple countries. The European 3G auctions were run sequentially (although those in Germany and the Netherlands occurred virtually at the same time). The pattern of prices in the European 3G auctions (table 7.1) suggests that these incentives were relevant in that early losers were discouraged from competing aggressively in the subsequent auctions.

Table 7.1:
European 3G auction prices

Country	Date	Per capita revenues
United Kingdom	March, April 2000	\$4.90
Germany	July, August 2000	\$4.74
Netherlands	July 2000	\$2.33
Italy	October 2000	\$1.84
Austria	November 2000	\$0.62
Switzerland	December 2000	\$0.13

One other form of complementarity arises when bidders have budget constraints: the more a bidder spends on one of two or more complements, the less money the bidder will have available for other lots. While, on the surface, the view that budgets can explain the pattern of European 3G auction prices appears plausible, this seems inconsistent with the bidding, unless bidders also were interested in aggregate coverage across Europe. There were 13 bidders that started in the UK 3G auction. The 8 losers did not spend any money, and presumably should have had money for subsequent auctions. In Germany, Debitel (partly owned by Swisscom) dropped out, but Debitel was not involved in the United Kingdom. So the 8 losers still should have had money to bid in the Netherlands, Italy, Austria, and Switzerland, as well as in the sealed-bid tender in France (where one of four 3G licenses went unclaimed at a minimum of 2 billion euros).

7.4 Sequential Intellectual Property Rights Auctions

Another area where sequential auctions occur is for new technology. Over time, technology improves. Often the firms developing the technology are not the ones that use it directly; rather, they license it to downstream competitors. Those owning the intellectual property (IP) can conduct competitive bidding processes to determine which firms will be able to use the technology. Often, as is assumed in this section, the IP owner will prefer to license a technology to one firm, rather than to multiple firms. This will be the maintained assumption in this section. The question addressed is whether the technology leader will tend to have an incentive to outbid rivals with inferior technologies. In other words, if a firm gains access to technology that is better (e.g., one generation ahead), possibly through an auction, then will it win an auction for rights to the IP for a next generation technology that represents an improvement over previous versions?

To capture these ideas, it is assumed that there are two firms, and each has access to a technology characterized by a single cost parameter. More specifically, it is assumed that a firm with generation n technology will have constant unit costs, c^n, and that $c^n > c^{n+1}$ for all n. It is assumed that there are no other costs.[5]

First, consider a simple two-period model. Each firms starts with a basic technology, with cost parameter c^0, available to any firm. Before each period there is an auction for the new technology. The first auction is for technology with costs c^1 and the second for technology $c^2 < c^1$. It is also assumed that technology 1 can be used for the second period before the auction for technology 2 is held, and that technology 2 will have a value derived only over the second period.

What follows derives the incentives of the technology leader to win the auction for technology 2. These incentives depend on the nature of the competition between the two firms. First consider the case where the firms compete in price, and there are perfect substitutes. If the leader wins the second auction, it will earn

$$\Pi^L = \max[D(p)(p - c^2) \mid p \le c^0],$$

where $D(p)$ represents market demand at price p. In contrast, the lagging firm will earn

$$\Pi^l = \max[D(p)(p - c^2) \mid p \le c^1].$$

Therefore $\Pi^l < \Pi^L$, so the leading firm will tend to win the second auction.

Now consider the case where there is a follow-up innovation that would achieve costs $c^3 < c^2$. Then the incentives of the leading firm to win are

$$\Pi^L = \max[D(p)(p - c^3) \mid p \le c^0],$$

and the lagging firm can offer as much as

$$\Pi^l = \max[D(p)(p - c^3) \mid p \le c^2]$$

in the second auction.[6] If there is a sequence of n non-drastic innovations,[7] that is,

$$\text{argmax}[D(p)(p - c^n) \mid p \le c^0] = c^0,$$

then a similar argument implies that the same firm would win each auction. It is possible to work backward and calculate the prices in each auction. To simplify exposition, it is assumed that $D(p) = D$ for all $p \le \overline{p}$.

The price in the last auction would be $[c^{n-1} - c^n]D$. In the next to last auction the price would be

$$D\{[c^{n-2} - c^{n-1}] + [c^{n-2} - c^n] - [c^{n-1} - c^n]\} = 2D[c^{n-2} - c^{n-1}].$$

A similar calculation can be used to establish the price of the mth-from-last auction as $mD[c^{n-m} - c^{n-m+1}]$. So, with equal cost improvements, namely $c^k - c^{k+1} = c^j - c^{j+1}$ for all j, k, the low-cost firm will always win, and prices will be decreasing from one auction to the next. This result has been called an *increasing dominance* result.[8] It is sensitive to the nature of the competition in the market for products or services using the technology.

To see this, consider the case where the firms are Cournot competitors with constant unit costs c that depend on the vintage of the technology, v, that each firm has. Suppose that one firm has a base technology c^0 and the other has a newer technology, $c^1 < c^0$, and that there is an auction for an improved technology $c^2 < c^1$. Suppose too that there is linear demand $P = a - bQ$, where Q is the total output of the two firms. Straightforward calculation shows that when the leading firm wins the auction for technology c^2, and the other firm has technology c^0, then the Cournot equilibrium quantities will be

$$q^L = \frac{a - 2c^2 + c^0}{3b}$$

for the leading firm, and

$$q^l = \frac{a - 2c^0 + c^2}{3b}$$

for the lagging firm. Profits will be

$$\frac{(a - 2c^2 + c^0)^2}{9b}$$

for the leading firm and

$$\frac{(a - 2c^0 + c^2)^2}{9b}$$

for the lagging firm. However, should the lagging firm win the auction for technology c^2, profits would be

$$\frac{(a - 2c^2 + c^1)^2}{9b}$$

for the firm that was previously the lagging firm, and

$$\frac{(a - 2c^1 + c^2)^2}{9b}$$

for the other firm. So the firm that starts with technology c^1 will want to offer

$$L = \frac{(a - 2c^2 + c^0)^2}{9b} - \frac{(a - 2c^1 + c^2)^2}{9b},$$

and the other firm will offer

$$l = \frac{(a - 2c^2 + c^1)^2}{9b} - \frac{(a - 2c^0 + c^2)^2}{9b}$$

to win technology c^2. Here L can be larger or smaller than l, depending on the values of c^0, c^1, and c^2. For example, if $a = 10$ and $b = 1/9$ and if $c^0 = 3$, $c^1 = 2$, and $c^2 = 1$, then $l > L$. But $L > l$ if $c^0 = 5$, $c^1 = 4$, and $c^2 = 3$.

7.5 Auctions for Complementary Inputs with Imperfect Competition in Downstream Markets

The preceding analysis also is useful in examining what happens when there is an auction for an upstream input that can affect downstream competition. Spectrum auctions are one example of this. Most spectrum auctions now are largely for incumbents competing to expand capacity and lower costs. The technology of spectrum is such that if one firm has more spectrum than its rival, it can offer service at a lower cost per unit of capacity—and thus with higher bandwidth and better quality.

To model this case, suppose that there are two firms that currently have costs c_j^0, $j = 1, 2$, and that there is an auction for a single block of spectrum that will reduce their marginal cost to $c^1 < \min\{c_1^0, c_2^0\}$. It is also assumed that, as in the previous subsection, the firms engage in Cournot competition, so that the price $P = a - bQ$, where Q is total output of the two firms.

The incremental profit for a firm that wins the additional block is

$$\Delta\Pi_i = \frac{(a - 2c^1 + c_j^0)^2}{9b} - \frac{(a - 2c_i^0 + c^1)^2}{9b}.$$

Whether the firm with lower initial costs will be willing to outbid its rival will depend on whether its incremental profit will be higher. With Cournot competition in the downstream market, it is possible that either the leading or the lagging firm can win, as in the previous subsection. However, if there is Edgeworth–Bertrand competition, the same firm will win each auction.

This issue is important in setting policy because spectrum is regulated by government agencies that often set caps to ensure that the market does not become excessively concentrated.[9] These types of models suggest that the market for wireless services can, under some conditions, become increasingly concentrated if there is a sequence of auctions that gradually release and reallocate spectrum over time. This section has merely served to introduce a framework for assessing the issue, rather than try to provide a complete analysis.

This section also serves to show that increasing dominance, and therefore concern about spectrum caps, can—in some, but not all, circumstances—be justified. The analysis of this issue is incomplete, and increasing dominance will be addressed in more detail in chapter 10.

The example above only considers sequential auctions. A regulator may have the option of selling blocks sequentially or in a single auction, and of allowing a single auction to include many small blocks or only a few (possibly only one) large ones. A simultaneous auction of all the spectrum blocks, or of multiple licenses for new technology, can have different outcomes, including the possibility of leapfrogging.

To see that leapfrogging may occur, consider the case where there is a single large block of spectrum available for bid, possibly instead of several small blocks. Also suppose that there are two firms bidding: a leader, L, and a lagging, smaller firm, s. Let $\Delta\pi_1^L > \Delta\pi_1^s > 0$ denote the incremental profits of the two firms. If the leading firm knows $\Delta\pi_1^s$, it will bid just a bit more than this and become more dominant. However, if the leading firm does not know $\Delta\pi_1^s$, then it will wish to maximize $[\Delta\pi_1^L - \Delta\pi_0^L - b^L] \times F(b^L)$, where $\Delta\pi_0^L$ is the leading firm's profit if its rival wins, b^L is what it bids, and $F(b^L)$ is the probability that this bid wins. This results in a first-order condition

$$\Delta\pi_1^L - \Delta\pi_0^L - b^L = \frac{F(b^L)}{F'(b^L)},$$

which means that leapfrogging can occur.

7.6 Summary

This chapter considered sequential auctions for complements. One striking result, which appears quite robust, is that the winner of a first auction for one of two or more complements does not really face an exposure problem. Admittedly, these models are highly simplified and abstract, and bidders do face risks that are not captured by them. However, the initial winner in the first auction will tend to have a strong advantage, which will tend to increase in a sequence of auctions for complements. The fact that an initial winner can lose a subsequent auction does not totally offset the benefits of winning an initial auction but may somewhat reduce incentives to bid beyond values initially in order to win the market.

A second striking result is that prices will not be uniform. The first auction will tend to attract price offers well beyond values, and subsequent auctions will see much less intense competition. So, for example, the initial award of a franchise or license can be very intense, but the re-auction, or renewal, is then likely to see much less or almost no competition. Prices in any one auction are not linked to underlying values for that auction, but to the entire sequence of auctions.

These issues arise in many auctions. For example, many auctions of licenses or franchises have limited durations, and do not necessarily specify all the renewal provisions in advance. The analysis here suggests that license durations that exceed the life of the investment may not be essential. However, the analysis provides a way of assessing the important role that the renewal provisions can take in determining bidder incentives and the outcome in the first of a series of auctions.

8 Single-Product Auctions

Summary

Many auctions include identical lots of a single product. Rather than auction each lot individually, an auctioneer can ask bidders to submit demand (or supply) curves to express how much each bidder wants at any given price. An alternative way for an auctioneer to elicit such demand (or supply) is in the form of a clock auction, in which the auctioneer names a monotone sequence of prices and asks bidders to indicate how much they want at each price. This chapter describes such auctions, and shows when a demand-function (or supply-function) game is strategically equivalent to a clock auction. It derives equilibria for demand- and supply-function games with and without uncertainty. It also explores the effect of capacity constraints. Lastly, it describes a number of applications of demand- and supply-function auctions and clock auctions.

8.1 Introduction

Often auctions include many units or lots of an identical product. Indeed virtually all auctions in the energy sector allow bidders to submit offers for one or more lots or blocks. The blocks can be 25 MW entitlements, rights to 100 MW or 1 MW of transmission rights, or other products. Multi-unit auctions are found in many other sectors. For example, the agricultural cooperatives Fonterra and Ocean Spray each run monthly auctions: In Fonterra's auctions bidders can buy any number of full container loads of various forms of milk products, and Ocean Spray's auctions are for cranberries and related products.

An auctioneer wishing to sell, or purchase, many units of a single product will need to decide whether to include all units in one bid, in a single auction, or in separate auctions. If multiple or all units are auctioned at one time, then the auctioneer needs to decide on what form bids should take and how to elicit information from bidders about how much they want at different prices. Bids can be demand functions, that is, separate prices for each unit. There can be restrictions on the form of demand function. The auction can allow bidders to submit one price for all units or different prices depending on the quantity bid. Alternatively, as in a clock auction described below, the auctioneer can announce a sequence of increasing prices and, at each price, ask each bidder to indicate its demand.[1] The auctioneer may also need to consider the potential impact of strategic withholding by bidders. This chapter provides a review of some of the more commonly used multi-unit auctions and describes their properties.

There are several ways in which multi-unit auctions are typically conducted. Probably the most common method for auctioning identical units of a product is by means of sequential auctions, and one at a time. This type of auction is easy to organize. But, as what follows should make clear, this may not always be the best method. Multiple auctions for each available unit can be held in parallel when the auctions are conducted online. The following discussion, and some examples from previous auctions, also suggest this is not usually the best way of organizing the sale of multiple units of a single good.

At times the auction format is influenced by indivisibilities in underlying production capacity. For example, in the energy sector, supplies are typically linked to generating units that have fixed capacities. Possibly for this reason many reverse energy auctions allow bidders to submit supply functions, and the form of the bid function that bidders are allowed to submit is based on generation facility capacities. The auctioneer then aggregates the supply-function bids, and determines the price and the allocation based on market clearing of the aggregate supply and requested demand.

The *clock auction* is another type of multi-unit auction in which all units are sold at one time. As is explained below, clock auctions are in some cases strategically equivalent to supply-function auctions. This strategic equivalence fails to hold when bidders are provided information about rival bids between clock auction rounds. The combinatorial clock auction (CCA) is a recently developed variant of the clock auction in which bidders can submit any number of package bids in a final sealed-bid round, and prices are based on a modified second-price rule.

This chapter examines auctions for multiple identical units of some product. The next section describes clock auctions. Clock auctions can differ in the information provided to bidders between rounds, and in the pricing rule. Clock auctions are compared with sequential auctions. The following section describes demand- and supply-function auctions under complete information. Then the outcome of clock auctions and demand-function auctions are compared. In all these multi-unit auctions, bidders' strategic withholding can adversely affect the outcome.

8.2 Simultaneous versus Sequential Auctions

The analysis of the last chapter indicates that the expected price in a sequence of auctions for identical products is a *martingale*; that is, the expected price in one auction equals the actual price in the previous auction, provided that bidders' valuations are the same across auctions at different times. This result no longer holds when bidder valuations can change. Should bidder valuations be static, then, as explained in the previous chapter, the price and allocation are independent of how many units are sold in each auction. Indeed the highest value bidders win, and all winners pay the same expected price.

When, however, bidder valuations can change , price can vary from unit to unit across auctions. What this means is that a higher value bidder can lose in an early auction to a lower value bidder, and fail to win later. In contrast, when bidder valuations do not change, this would not occur. A high-value bidder can, in theory, lose to a lower value bidder due to a possible miscalculation by the losing higher value bidder or because the high-value bidder underestimates the level of competition in the later auctions.

Unexpected fluctuations in demand across auctions can be difficult to plan for. Even with little or no fluctuation, high-value bidders can lose to lower value bidders in sequential auctions. This type of outcome reduces the average bidder surplus as well as overall auction revenues. Thus sequential auctions require bidders to make complex calculations and forecast future demands. As is explained below, in simultaneous clock auctions such guesses are not necessary, eliminating the potential for mistakes.

Nevertheless, sequential auctions do have advantages when demand across bidders can shift over time. As explained in the previous chapter, overall auctioneer revenues (for a forward auction) will tend to be greater when the auction volume is split across two or more auctions.

If this benefit is small, then the possible inefficiencies due to bidder miscalculations and bidding errors can offset the benefits of sequential auctions.

8.3 Clock Auctions

A clock auction is probably the simplest multi-unit auction. In a clock auction the price starts below market-clearing levels, and the auctioneer raises prices in a series of small discrete steps. After each upward tick of price, bidders indicate how many units they want to purchase. Prices keep rising until there is no longer any excess demand. When the clock stops, the remaining bidders are the winners, and each bidder is usually asked to pay the last posted price. Other pricing rules and stopping rules can be used. For example, all bidders can be required to pay the last price for which there is excess demand. Another alternative is that the clock price increases until there is no longer *any* demand, and each bidder can be required to pay the last price before it drops out.

The clock auction is essentially a simple adaptation of the classic Walrasian tâtonnement process.[2] The main distinction between the clock auction and the Walrasian auction is that the supply side is absolutely inelastic in the clock auction but makes simultaneous offers in the Walrasian auction. The set of clock round offers traces out a demand schedule. The auction thus ends when the price rises to the point where demand drops to or below supply. In the remainder of this section, I briefly describe a few properties of the clock auction and a few variants.[3]

The simplest case is where each bidder will purchase at most one unit. In this case each bidder should remain in the auction until the price reaches its value.[4] If a bidder were to drop out before the price reached its value, it would derive no benefit from doing so—and this can only save rivals money. There is similarly no point in waiting until after price goes past value, as the bidder may then overpay. The bid strategy is essentially the same as in a standard English auction.

On one hand, if bidders pay the last price for which there is excess demand, then it is a dominant strategy to bid up to value and then drop out. In other words, bidders will want to bid value, independently of rivals' values and bids. On the other hand, if bidders pay the last price posted, which means the first price posted for which there is no longer excess demand, then it is possible for a bidder's offer to affect how much it pays. In this case a bidder's strategy can depend on the tie-breaking and rollback provisions. Specifically, if at the next to last price demand exceeds supply, but at the last price there is excess supply (i.e., there is overshooting), then a bidder may want to drop out a bit before the price reaches its value. This is especially true if the increments are large or the bidder believes that overshooting is likely.

Thus the rule that determines prices can have some effect on the bidders' strategy. If bidders pay the highest losing offer, then bidding value is a dominant strategy. If bidders may have to pay the lowest winning bid, then they can have incentives to drop out early. Note that these differences are likely to be small if there are a large number of bidders and values are closely bunched, that is, if the gap between the highest losing offer and the lowest winning offer is small.

8.3.1 Clock Auctions and Sealed-Bid Auctions

The decision facing a bidder that is seeking to purchase up to one unit in a clock auction is very similar to the decision facing a bidder in a sealed-bid auction. Indeed at times the decisions will be strategically equivalent. When this is the case, the outcomes of the two types of auctions will be the same.

The only differences between a clock auction and a sealed-bid auction are in the information available to bidders when having to decide on final bids. In the extreme case of a clock auction in which the auctioneer (1) never reports any information about excess demand and (2) increases price until there is zero *demand*, and not just zero *excess* demand, the decision facing a bidder is identical in a second-price sealed-bid auction and in a clock auction in which the bidder pays the high losing bid. A bidder has no information about rivals' participation in the clock auction. And so, each bidder is just choosing a price at which to drop out of the auction. This stopping price should equal the bidder's value, and be exactly the same as in the second-price sealed-bid auction.

Similarly the strategic decision facing a bidder in a clock auction is the same as in a sealed-bid auction when (1) the clock auction price is the last amount that the bidder was willing to offer and (2) the bidder pays its bid amount in the sealed-bid auction. In both cases the bidder will want to shade its bid, that is, drop out below value. The less it bids, the greater is its surplus, but the lower the chance of winning. And, as in the clock auction, the bidder gets absolutely no information about rivals, so its choice of when to drop out is just based on when it thinks the next highest value rival would drop out.

This strategic equivalence of clock auctions and sealed-bid auctions breaks down when information is reported to bidders between price increments (rounds). When, in a clock auction, bidders pay the highest losing bid, information about rival demand does not affect strategy except to the extent that this information affects bidder valuations. This means that bidders having affiliated values will tend to bid more aggressively in the clock auction than in the sealed-bid auction.

If the clock auction price is uniform and set at the lowest winning bid, then the strategic equivalence of sealed-bid auctions and clock auctions breaks down, even with independent private values—as least when a bidder has at least two remaining rivals its ante prior beliefs are that the likelihood of ties if very low, or when increments are small. When bid increments are very small a bidder will know that there is no excess demand, and will not stop bidding as long as there is any excess demand and the price is still below value. Thus, in a clock auction, bidders won't always need to bid up to the point where the benefit of increasing the probability of winning by offering more just offsets the reduction in surplus. Often the auction will end sooner. The auction can also end later, if a bidder sees competition staying in longer. Thus there can be some gain to having the additional information in the clock auction. It is possible that a bidder will choose the same price at which to stop bidding in a clock auction as it would offer in a first-price auction, but this would depend on the realized distribution of bidder valuations. Klemperer (2004) shows that expected revenues will be the same in the case of independent private values. The RET does not apply, though, in cases of affiliated values: the winning bidder need not have to pay as much as it would offer in a first-price auction, but that bidder may also bid more aggressively due to having more information available.

8.3.2 Clock Auctions, Sealed-Bid Auctions, and Market Power

We now consider the case where a bidder may want to purchase more than one unit. In this case its decision to reduce demand can reduce the price it pays for the units it does purchase. In a *second-price* auction, that is, an auction in which winning bidders all pay the highest losing bid amount and each bidder wants to purchase at most one unit, it is a dominant strategy for each bidder to bid its value. However, if some bidder is seeking to purchase two or more units, it can have an incentive to bid less than its value in some cases.

A simple example can illustrate the incentives to withhold demand in this case. Suppose, to start, that there are three bidders, all values are drawn from a uniform distribution on [0, 1], and that there are two units for sale. One bidder has a demand for two lots, and the other two bidders only want one lot each. The top two bids win and all winning bidders pay the third highest bid amount. The expected values of the four units demand are, in rank order, $\frac{4}{5}$, $\frac{3}{5}$, $\frac{2}{5}$, and $\frac{1}{5}$. Assuming that the large bidder does not shade its bids for either of the two units it is offering to purchase, the expected price is $\frac{2}{5}$.

Now consider the incentives of the bidder seeking two units to shade its bid. The possible rank orders (where *L* denotes the large, two-unit bidder, and *s* one of the two small, one-unit bidders) are *LLss* (i.e., the two-unit bidder *L* has the highest two valuations, and the other two bidders *s* have the lowest valuations), *LsLs* (i.e., *L* has the highest and third highest valuations), *LssL, sLsL, sLLs,* and *ssLL.*

In the case *LLss*, the two-unit bidder will win two blocks with an expected price of $\frac{2}{5}$. This occurs with probability $\frac{1}{12}$. So this bidder, by bidding 0 for its second unit, loses, and loses an expected payoff of $\frac{1}{5}\times\frac{1}{12}$. However, bidding 0 for the second unit will allow this bidder to pay on average $\frac{1}{5}$ instead of $\frac{2}{5}$ in the cases *LsLs*, which occurs with probability $\frac{1}{6}$, and *sLLs*, which occurs with probability $\frac{1}{12}$. In the other three cases, underbidding does not affect this bidder's expected payoff. So this bidder gains an expected net return of $\frac{1}{6}\times\frac{1}{5}=\frac{1}{30}$ by strategic underbidding. This type of strategic demand reduction has been noted as possible both in spectrum auctions[5] and in the energy sector.[6]

Note that if bidders pay the lowest winning bid, then the outcome will generally be the same as in the case where bidders pay the highest losing bid, assuming that each bidder is seeking to win only one unit. When bidders are seeking multiple units, they will also have some incentive to underbid.

8.4 Supply-Function Auctions

This section considers reverse auctions in which bids are in the form of supply schedules. More specifically, each bidder informs the auctioneer how much it is willing to supply as a function of the price it is paid. Thus the bidder may offer a quantity q_0 at price p_0 and a quantity $q_1 \geq q_0$ at price $p_1 > p_0$. These supply-function bids are a common way for power to be offered in energy auctions.[7] The reason for the use of supply functions is on the technical side: the offers can be matched to the characteristics of the plants providing energy[8] (although, in many cases, energy offers are in the form of financial contracts—the supplier guarantees a spot price determined by a grid operator). As Green and Newbery (1992) noted, the reliance on supply-function bids may have the unintended consequences of resulting in higher (i.e., less competitive) transaction prices.

A very simple example of a supply-function game is one in which two plant owners, say having $M = M_1 = M_2$ MW of capacity, can offer to sell it in to the grid operator a price p_j for operator $j = 1, 2$. Suppose too that the needed supply, S, satisfies

$$\frac{M}{2} < S < M.$$

For example let $S = 100$ and $M = 60$. Also assume that the grid operator, who serves as the auctioneer, will cover the fuel costs, and so bids are just for rights to use the capacity. The grid operator is assumed to have a reserve price R, which for purposes of this example we take to be 100. It is further assumed that operators receive their bid price.

It is clear that no operator would ever accept a price less than

$$\frac{S-M}{M} \times R,$$

or $66\frac{2}{3}$ in this example, the reason being that the bidder can always sell $S - M$ at the reservation price. However, when one bidder chooses a price strictly above

$$\frac{S-M}{M} \times R,$$

its rival will have incentives to undercut. Thus there is no pure strategy equilibrium. It turns out that choosing a probability distribution function $F(p)$ such that each rival earns the same expected profit, independent of the price the rivals choose between

$$\frac{S-M}{M} \times R$$

and R, is a mixed strategy equilibrium. This is the case if

$$Mp[1 - F(p)] + (S - M)pF(p) = k$$

or

$$F(p) = \frac{M - (k/p)}{2M - S}.$$

It can also been seen that

$$k = M\left[\frac{S - M}{M} \times R\right].$$

In the numerical example

$$F(p) = \frac{60 - 4{,}000/p}{20}$$

for $66.67 \leq p \leq 100$.

More generally, consider the case of two firms bidding into a market (auctioneer) demand function $Q = D(p)$.[9] Let $S^j(p)$ denote the supply schedule offered by firm $j = 1, 2$. What follows is a description of some of the necessary conditions for an equilibrium set of offers $S^{*1}(p)$, $S^{*2}(p)$ and resulting equilibrium price p^* and quantity Q^*. The assumption that the auctioneer incurs all variable costs, as is often the case in energy markets, is maintained.

Note that if there is an equilibrium, unique or otherwise, then $D(p^*) = S^{*1}(p^*) + S^{*2}(p^*)$; that is, supply must equal demand. If there is no equilibrium, it is assumed that neither firm earns any profit. Now consider *any* pair of firm quantities (q^1, q^2), such that the price p satisfies

$$p = D^{-1}(q^1 + q^2).$$

What follows derives supply functions for the each of the two firms that can support these prices as an equilibrium outcome. Notice that given $S^i(p)$, firm j will want to maximize $p[D(p) - S^i(p)]$ with respect to p. This implies that the following first-order conditions need to be satisfied:

$$[D(p) - S^i(p)] + p[D'(p) - S^{i\prime}(p)] = 0. \tag{8.1}$$

But (8.1) is satisfied whenever[10]

$$S^{i\prime}(p) = D'(p) + \frac{D(p) - S^i(p)}{p} = D'(p) + \frac{q^j}{p}. \tag{8.2}$$

However, multiple values of the price will satisfy (8.2). Thus there will be multiple equilibria. For example, if demand is linear [$D(p) = A - Bp$], then linear supply schedules $S(p) = s \times p$ will satisfy (8.2)[11] whenever

$$s = \frac{A}{p} - 2B.$$

Next consider the case in which there is uncertainty. For example, suppose that $D(p) = A - Bp + \varepsilon$, where ε is a nondegenerate random variable; that is, the demand has a random component. In this case the supply schedule each bidder chooses has to satisfy the first-order conditions in (8.1) for any realization of the random variable. This means that a pair of supply functions that are an equilibrium for one particular realization of demand need not still be one for others. It can be shown that there will be a unique pair of supply functions that will be an equilibrium for all realizations of demand.

8.5 Conclusions and Applications

This chapter has reviewed clock auctions and supply-function auctions. The two are very similar and, as explained in section 8.3, are identical when there is no information reported to bidders until bids go to zero in the clock auction. When bidders see when rivals drop out, or have some aggregate measure of rival offers, then the clock auction allows bidders to bid a bit more aggressively than in a sealed-bid supply-function auction.

In practice, supply-function (or demand-function) auctions are common in the energy sector, and in some other sectors. There are often restrictions on the form of bids; for example, bidders can submit only step function bids or piecewise linear schedules. These restrictions tend to reduce the complexity of the bidding process.

A key issue is how well clock and supply-function auctions work. Note that bidders can have market power in these auctions. This is most clearly seen in the simple capacity example of section 8.3.2, where each of two bidders offers its capacity to a grid operator. This result can be generalized; it can be shown that in cases where bidders can be pivotal, bidders' offers will always be bounded away from their competitive levels. This possibility was noted in Green and Newbery (1992), seen in practice, and proved in theory.[12]

The logical question is what are the alternatives to these types of auctions. One of the key issues in auction performance in multi-unit auctions is the incentives of bidders to make offers close to marginal values. The auctioneer can choose to reduce volume if offers are insufficiently competitive, and this can serve to elicit more competitive auctions from bidders.[13] However, this may only delay the inevitable. If bidders know that the auctioneer has to conduct another auction for what is left over, they will want to wait for the re-auction. On the one hand, if the conditions of Weber's martingale theorem hold, that is, bidders' values are static, then the re-auction will not help. On the other hand, if the bidders' values can increase or decrease, then, as explained in chapter 6, the auctioneer would be well served by reserving the option of deferring some purchases or sales to a later auction, and making this decision endogenous, that is, based on the bids.

9 Simultaneous Auctions

Summary

A novel approach to auctioning multiple objects began when the US Federal Communications Commission (FCC) conducted the first simultaneous ascending multiple-round (SMR) auction in 1994. This approach allowed bidders to link their offers for different objects—they could arbitrage substitutes and limit the risk of winning one of a set of complementary objects in SMR auctions. In other words, the SMR auction was probably the first type of auction that allowed bidders to make simultaneous decisions for a set of objects whose values were related. Prior to this, these decisions had to be made independently, creating all sorts of difficult decision problems for bidders. The technology for auctioning multiple objects with interdependent values is still improving. This chapter provides an introduction to the theory and practice of multi-object auctions.

9.1 The SMR Auction

It might not be hyperbole to state that the adoption of the simultaneous multiple-round (SMR) auction by the US Federal Communications Commission (FCC) was the single most significant event in the application of game theory to address practical decision problems. This auction format was developed de novo to help the FCC, a governmental agency, more efficiently allocate an extraordinarily valuable resource, generating over $10 billion in nominal auction revenues in the first few years. The problem facing the FCC was to develop an auction mechanism that would efficiently allocate spectrum licenses, that is, licenses to use specific frequency channels over different geographic areas. The licenses could be substitutes, as where there are two or more similar or identical blocks covering the same geographic areas, or complements, as where a bidder is seeking different bands covering the same area for different needs, or in different areas (where coverage is an issue).

This allocation presented several challenges. One is that in parallel or sequential auctions substitutes for identical products often sell for different prices. In such cases a losing bidder for one lot can have a higher willingness to pay than a winning bidder for an identical lot. This also means that the outcome of the auctions is necessarily inefficient. Reallocation through after-market transactions cannot be relied upon to correct for such inefficiencies unless transaction costs are low. This same logic extends to close but imperfect substitutes; it is possible that a higher value lot will sell for less than a lower value one.

Another challenge is the *exposure* problem: a bidder entering separate bids for two or more complements may end up overpaying if it wins part, but not all, of its desired package. For example, if one block is worth 10, but two blocks are worth 50, a bidder that offers 20 per block may win only one. Allowing bidders to submit all-or-nothing package bids is one approach to this exposure problem. However, package bidding introduces other complexities, as is explained in more detail in chapter 10. And auction rules that help one bidder address the exposure problem can create a bias in favor of a large package bidder over bidders seeking smaller packages.

The chapter starts with a brief description of the SMR and clock auction formats. A simultaneous clock auction differs from an SMR auction in that bidders name only quantities based on prices set by the auctioneer in each round. This chapter reviews properties of the SMR auction format. Then it provides a review of the experience with SMR and simultaneous clock auctions. In particular, it presents both good and bad experience with the SMR auction. Finally, it explains how the performance of an SMR auction (i.e., how close it comes to achieving an efficient outcome) and the revenue raised can be surprisingly sensitive to seemingly minor decisions about auction configuration and parameter settings. Some guidance is provided about how to ensure that an auction outcome is more likely to be efficient, and to counter bidder efforts to exert market power.

9.2 Introduction to the SMR Auction

9.2.1 A Brief History of the SMR and Clock Auctions

In 1994 the US FCC faced a congressional mandate to allocate and assign various combinations of regional and national spectrum licenses using auctions.[1] Because of incumbent users on the spectrum, the new licenses would, in some cases, be awarded to firms competing with existing licensees, and in other cases, existing licensees would be seeking to add to their frequency holdings. Additionally both bandwidth and geographic efficiencies meant that in some instances a pair of licenses covering the same or adjoining regions could be substitutes or complements. The FCC's mandate included efficiency and the promotion of deployment of advanced telecommunications services as primary goals, with revenue a secondary goal.[2]

Thus the FCC faced a difficult auction design problem—it wanted an auction that would efficiently allocate licenses that could be complements or substitutes. This meant that the auction should allow efficient arbitrage in the case of substitutes. And, in the case of complements, the auction should allow bidders to effectively manage the exposure problem when bidding for a package of complementary licenses. Moreover, the FCC would not want to allow bidders to win too much spectrum so as to create market power. An additional complication was that there were different options for dividing the spectrum bandwidth with respect to geographic partitioning of the United States and its territories.

9.2.2 SMR Auction and its Variants

This section provides a very brief description of the basic SMR and simultaneous clock auction rules. As noted above, the first SMR auctions, introduced by the US Federal Communications Commission, were initially introduced for the sale of a number of lots, some of which could be substitutes and some could be complements in a single auction. SMR and clock auctions are intended for selling multiple units or blocks in a single auction. In an SMR or clock auction bidding occurs in a sequence of rounds.

Prior to the start of an SMR or clock auction, the auctioneer posts some application and eligibility requirements. One component of the initial eligibility requirements includes the application for eligibility points; the points per lot are based on the scale of activity or set by the auctioneer for what is available. Bidders must include in their application some indication of how many points they want to be able to bid for. A bidder is never allowed to win more points than it indicated in its application. This is a component of auction rules designed to encourage increasingly serious bids as the auction progresses.

In the first round of any SMR or clock auction, a bidder is allowed to enter bids for any combination of lots whose total activity points do not exceed that bidder's initial eligibility. If a bidder fails to meet its *activity target*, which is some fraction, up to 100 percent, of its eligibility, it will see its eligibility decrease in the next round. For example, if a bidder's eligibility is 200 activity points, the required activity percentage is 80 percent, and this bidder is active on 64 points in the first round, its eligibility for round 2 will fall to 80 = 64/80 percent. And once a bidder loses eligibility, it can never regain it.

After the first round and each subsequent round, the auctioneer will post—possibly partial—information about the bids in the previous round. At a minimum, the auctioneer will publish the minimum required bids for the following round, and usually some measure of the excess demand for each lot. In most of the first SMR auctions, the auctioneer would report all bids and bid amounts.[3] Also, after each round, the auctioneer informs each bidder of the lots for which it has a provisional winning bid.

In contrast, in many SMR auctions, a bidder can submit jump bids, that is, bids above the minimum required amounts. This can serve to influence the rate at which prices of different lots increase during the auction, as well as the relative prices. A bidder may wish to do so when there are complements and the bidder seeks to determine its ability to win one lot before having to commit to another.

In each round after the first, bidders that were topped on any lot can improve their offer, switch their points to other lots, or reduce eligibility. A bidder failing to maintain its required activity in one round will lose eligibility in the next. Note that typically a bidder does not have to increase its offer on a provisional winning bid to have that bid count toward activity.[4] As prices rise,[5] bidders will find the cost of keeping eligibility increases, and so will have incentive to reduce activity, and eligibility. Also, if the auction starts with less than a 100 percent activity requirement, the activity requirement will tend to increase toward 100 percent. So, as the auction progresses, bidders will tend to drop points and demand. The auction ends when there is no longer excess demand on any lot.

There are many variants of the SMR auction format. Perhaps the most fundamental variant is the simultaneous ascending clock auction, in which bidders do not name prices, and only indicate which lots they want at the announced clock prices. As there is often little incentive for bidders ever to bid much or anything above minimum required bids, the simultaneous clock auction[6] is really not much different than the SMR. The inability of bidders to submit jump bids, that is, increase offers above the minimum required bids, has two effects. One is that there is less scope for signaling in the clock auction than in an SMR auction. The other is that relative prices are controlled by the auctioneer. The lack of signaling can produce more competitive outcomes. The fact that bidders have less influence on relative prices in a clock auction means that decisions may be sequential—the activity rules require decisions on large lots to occur before decisions on smaller ones, making the auction sequential, which can result in inefficiencies and price anomalies.

SMR auctions vary in many other respects. As noted above, the activity requirements and information reported to bidders are auction design features. The effect that these design features can have on the auction outcome is explained below. SMR and clock auctions can be run for multiple units of a single product, single units of multiple products, and multiple units of multiple products. Simultaneous auctions have also been used for procurement.

9.3 SMR Auction Properties

The SMR auction is quite a bit more complicated than a standard English auction or a sealed-bid auction. The question arises as to what value is there in using an SMR or a clock auction format. The SMR auction can, under circumstances described in what follows, effectively allow bidders to arbitrage price differentials. This ensures that the outcome is efficient, unlike the case in sequential auctions.

9.3.1 Straightforward Bidding in SMR Auctions

As has been discussed above, sequential auctions, or simultaneous auctions for substitutes or complements, can result in difficult decisions for bidders and in inefficiencies. The SMR auction gives bidders the ability to arbitrage price differences between substitute products. In other words, a bidder can shift activity from lots for which the value-price differential is low to one for which it is high. Thus, a bidder can bid *straightforwardly*. In other words, bidders can always bid on the set licenses for which the differences between combined values and prices is the largest. Note that this requires that bid increments be small, and the activity rule provides flexibility to allow arbitrage. For example, suppose lots A and B are very close substitutes for a bidder, but that lot A has more activity points than B. If the prices are about the same, a bidder may prefer A to B.

If price of A is a fixed amount, Δ, more than B, the bidder would prefer B. If the prices start the same, this bidder may start bidding for lot A. However, if the price of A rises faster than that of B, the bidder would want to and can switch to B. Now if later in the auction the price of B rises to catch up with, and perhaps pass, A, this bidder would want to switch back. But an activity requirement of 100 percent (or more than the ratio of the points for B to the points for A) would not permit this.

When such arbitrage can occur during the SMR auction bidding rounds, then, on the margin, a bidder will be indifferent between two lots only when the price-value differential is the same across the two. As all bidders can shift, the auction should end where the marginal bidder is just indifferent. This is a standard criterion for efficiency in competitive markets. More specifically, assuming bidders always bid on the set of blocks that maximize surplus in each round, the outcome of an SMR or clock auction should be both efficient and a *competitive equilibrium*. In other words, at the final prices, excess demand is zero for all lots, which is the criterion for a set of prices to be a competitive equilibrium.

Summarizing, this logic implies the following result (Milgrom 2004, p. 270), which applies to a simplified SMR auction format in which bidders continue to bid on all they want until they have decided to drop eligibility in each round, and in which activity rules allow switching consistent with revealed preferences.

Theorem 3 *Straightforward bidding is a feasible strategy for bidder j for all initial prices, all fixed increments, and all feasible price paths iff all goods are substitutes for bidder j.*

This theorem does not apply when, as discussed above, the auctioneer does not assign close substitutes the same number of activity points, thus limiting bidders' opportunities for arbitrage. Note too that if goods can be complements, and there are two packages, and a bidder gets topped on some, but not all, lots in a package, then it may lack the eligibility to switch to what is a higher surplus package at the current round's prices. This can happen when, for instance, prices on the lots in one package increase, reducing the surplus that bidders can derive from it, while prices in another package do not.

The fact that bidding converges to a competitive equilibrium, which is also an optimum, follows from the fact that straightforward bidding means all bidders are able to arbitrage price differentials. So, if one bidder valued one unit of one product more than a rival who won it, it would always have the opportunity to outbid that rival, and so the auction could not end with a suboptimal allocation, and the prices could never be other than those corresponding to a competitive equilibrium. Summarizing:

Theorem 4 (Milgrom 2004) *Assume that all goods are substitutes for all bidders and that all bidders bid straightforwardly. Then the auction ends within a finite number of rounds. The final provisional winning bids and allocation of goods are a competitive equilibrium for an economy in which bidder valuations are within some small $\varepsilon > 0$ of the bidders' true valuations for all bidders. The final assignment maximizes total welfare within a single bid increment.*

Notice too that an SMR auction must converge in a finite number of rounds, assuming fixed increments. This follows from the fact that there is only room for a finite number of rounds before all prices would exceed all bidders' values. Thus, given any initial prices, any SMR auction converges to some limit. Further, given the arbitrage condition with substitutes, this limit must be a competitive equilibrium, assuming all lots are substitutes and bidders bid straightforwardly. Thus, a competitive equilibrium always exists for the case of substitutes.

9.3.2 Strategic Withholding in SMR Auctions

The preceding subsection shows how SMR and simultaneous clock auctions are efficient when bidding is straightforward and all the lots are substitutes. This subsection looks at the incentives for straightforward bidding and suggests some measures that can improve those incentives.

To illustrate why an individual bidder may not always want to bid straightforwardly, consider an auction in which two bidders are competing for two lots. The first bidder has a valuation for two lots, and the second bidder only wants one. Other bidders are assumed to have lower valuations. Suppose when prices rise to 50, this two-lot bidder, who values each lot at say 100 or more, sees it can win one lot. If this bidder believes it can win both lots for a price not much higher than 50, then it will want to keep bidding. However, if this bidder may have to bid nearly 100 to win both lots, or may not be able to win the second lot at all, then it will want to drop its demand for the second lot as soon as it sees it can lock down the first lot. Even if this two-lot bidder believes it is just as likely to be able to win the second lot at 50 as at

100, it will have no incentive to continue bidding. In other words, this bidder will wish to withhold demand.

This logic readily extends to situations commonly encountered in SMR spectrum auctions in which two bidders are competing for multiple blocks or in several geographic areas. If each were to accommodate the other, then the auction could end at a low price; on the other hand, should they keep bidding, the auction could end at a high price for the same allocation that might have occurred had the two each reduced demand. Early SMR auctions allowed bidders to submit arbitrary bid amounts. So a bidder could submit a very large bid, and add a few dollars or euros to bids on some lots over others. This type of bidding, that is, adding a few "trailing digits," can be interpreted as conveying signals.[7]

Regulatory authorities now usually specify a limited set of allowed bid increments. Thus, in some auctions, a bidder can only bid a fixed increment, and in other auctions a bidder can choose among a fixed set of increments (e.g., 10,000, 20,000, ..., 50,000 above the provisional winning bid. These nondiscretionary increments limit the possibility of signaling. Of course, overt collusive agreements among bidders are never allowed. So, in the end, signals are never more than cheap talk. But the cheap talk can, at times, affect the outcome.

Moreover, by not reporting the exact demand after each round, the auctioneer can further reduce bidder incentives to withhold demand. In the first example, if the two-lot bidder does not know that there is exactly one lot of excess demand, it cannot know that by reducing demand it can end the auction. Thus it is now common for auctioneers to limit the information reported between rounds about excess demand.

These limited-disclosure rules can come at a cost. Recall that in a common-value auction, aggregate information about rivals' demand can reduce uncertainty about a bidder's own value forecasts. This reduction in uncertainty can result in higher auction revenues. Thus, an auctioneer risks reducing auction revenues when it reports no information about excess demand to any bidders between rounds. At times, auctioneers will report approximate excess demand. For example, rather than indicating that there is only one block of excess demand, the auctioneer can indicate that there are fewer than five blocks of excess demand. Such information can alleviate the winner's curse in common-value auctions. Indeed, if the information reported is sufficient,[8] in the sense that any additional information would not help bidders improve their value forecasts, then full disclosure will have no additional benefits. In such cases it may be possible to provide adequate information to bidders to take full account of the winner's curse without having to risk increasing incentives for strategic withholding.

Along these lines is the decision that the FCC and other regulators have taken to limit bidder discretion in setting bid increments. In most of the early SMR auctions, bidders were allowed to submit bids in any amount, and the last few digits of bids may have been used in efforts at covert signaling.[9] In one case signaling resulted in a post-auction conviction of a bidder who attempted to collude during an auction with another bidder. Most recent auctions allow a limited number of bid increments. In some cases the auctioneer allows only yes or no bids for a fixed increment, which turns the auction into a simultaneous clock auction.

One other approach has been used to offset bidder incentives to strategically withhold demand. In the example above, in which there are two lots available in the auction, the two-lot bidder will have less incentive to withhold demand if the auctioneer can decide to only sell one lot. For example, the auctioneer may decide, in advance, to sell only one lot to any single bidder unless that bidder is willing to pay at least 75 or at least 50 percent more than a rival is willing to pay for the second lot. In this case the large bidder can still withhold demand, and still have an incentive to do so, but its incentive is reduced.[10] This approach can be effective when the auctioneer does not have to sell its entire inventory, and can carry it over to another date. It has been used in energy procurement auctions.[11] When the auctioneer does not have the ability to carry over the inventory to another date, or other outlets, then this type of volume reduction will not be effective. In some cases there may not even be the option of just keeping the unsold inventory.

9.4 SMR Auction Experience

The first few auctions were intended to assign rights to narrowband and broadband personal communications services (PCS) licenses. Narrowband PCS was intended for one-way and two-way paging and messaging services. The broadband PCS was for first digital (or 2G) mobile voice and data service. One part of the auction process was to determine the geographic area of coverage and the bandwidth associated with each license. In the first narrowband PCS auction, the FCC divided up the narrowband spectrum into ten national licenses of different bandwidths. In the second, it divided the spectrum into six bands in each of five geographic regions. The broadband PCS was initially scheduled to occur in a sequence of three auctions. The first, FCC auction 4, included two 30 MHz licenses in each of 51 *major trading areas* (MTAs). The second, for designated entities (small businesses, minority- and women-owned businesses, and rural carriers), was for another 30 MHz license in each of 493 *basic trading areas* (BTAs), which were finer than the MTAs. The third auction was for three 10 MHz blocks in each of the 493 BTAs.

The very first SMR auction—FCC auction 1, table 9.1, for data communications services—appeared quite successful in that it apparently facilitated arbitrage. The frequencies were divided into three types—five two-way licenses with 50 kHz uplink and downlinks, three two-way licenses with asymmetric 50 kHz downlink channels and 25 kHz uplink channels, and two one-way 50 kHz licenses. Each license with a particular bandwidth sold for essentially the same price as every other license of that type.

The second SMR auction (table 9.2)—(which was actually the 3*rd* FCC auction[12]—included six licenses in each of five regions. Again, the outcome strongly suggests that the SMR auction is an effective mechanism for assigning spectrum licenses. Note that in this case the regional blocks were likely complements. The activity rules apparently provided bidders adequate flexibility to solve their exposure problems.[13]

Table 9.1
FCC auction 1: national narrowband PCS

Name	Type	Final bid	Minority credit	Winner	Round
N-1	50–50	80,000,000	0	Pagenet	37
N-2	50–50	80,000,000	0	Pagenet	37
N-3	50–50	80,000,000	0	McCaw	33
N-4	50–50	80,000,000	0	McCaw	33
N-5	50–50	80,000,000	25%	MTel	37
N-6	50–12.5	47,001,001	0	AirTouch	24
N-7	50–12.5	47,505,673	0	BellSouth	25
N-8	50–12.5	47,500,000	25%	MTel	24
N-10	50–0	37,000,000	0	Pagenet	45
N-11	50–0	38,000,000	25%	Pagemart	46
Total		**671,006,674**			

Table 9.2
FCC auction 3: regional narrowband PCS

Band/ region	Northeast	South	Midwest	Central	Western	US total
50 × 50 kHz	$17,500,000	$18,400,000	$16,810,000	$17,340,000	$22,549,020	$92,599,020
50 × 50 kHz*	$14,850,000	$18,780,000	$17,360,401	$17,136,000	$22,800,000	$90,926,401
50 × 12.5 kHz	$9,471,082	$11,800,007	$9,291,000	$8,250,000	$14,857,003	$53,669,092
50 × 12.5 kHz	$8,949,543	$11,543,007	$10,057,004	$8,791,001	$14,281,111	$53,621,666
50 × 12.5 kHz	$8,675,000	$8,000,013	$9,500,000	$8,262,000	$14,281,001	$48,718,014
50 × 12.5 kHz*	$10,251,000	$11,262,003	$10,251,001	$10,488,000	$10,920,600	$53,172,604

Table 9.3
UK 3G auction results

License	High bidder	Final price £m
A	TIW(Hutchison)	4,348.70
B	Vodafone	5,964.00
C	BT3G	4,030.10
D	One2One	4,003.60
E	Orange	4,095.00

The SMR auction format has been used by regulatory authorities for a great many auctions for spectrum and other products and assets across the world since the mid-1990s. For the first wave of European spectrum auctions, the SMR format was used in Austria, Germany, Italy, the Netherlands, Switzerland, and the United Kingdom. In virtually every case, similar licenses sold for similar prices, and larger licenses sold for more than smaller ones. All these auctions included 120 MHz of paired UMTS spectrum, which could be divided into 4, 5, 6, or 12 lots.

The United Kingdom and Italy decided to divide the spectrum into three lots of 20 MHz and two lots of 30 MHz. One of the larger lots was reserved for an entrant.[14]

The United Kingdom was the first European country to conduct a 3G auction. That auction attracted the four incumbents (BT, Vodafone,

Table 9.4
German 3G auction

Blocks	Bidder name	DM	USD
1	Viag Interkom	8.3104 billion	3.86 billion
2	Mobilcom Multimedia	8.17 billion	3.79 billion
3	Mannesmann Mobilfunk	8.33 billion	3.87 billion
4	Group 3G	8.3046 billion	3.85 billion
5	Mobilcom Multimedia	8.2 billion	3.81 billion
6	Viag Interkom	8.2066 billion	3.81 billion
7	T-Mobil	8.3043 billion	3.85 billion
8	E-Plus Hutchison	8.2743 billion	3.84 billion
9	T-Mobil	8.2779 billion	3.84 billion
10	E-Plus Hutchison	8.1439 billion	3.78 billion
11	Mannesmann Mobilfunk	8.1438 billion	3.78 billion
12	Group 3G	8.1414 billion	3.78 billion

Orange, and One2One) and nine potential entrants. Each firm could bid for one license and had to remain active to remain in the auction. The four incumbents each won a license (and Vodafone won the larger, 30 MHz license). H3G won the license reserved for the entrant. The auction results were as in table 9.3.

The German 3G auction was designed differently. The same 120 MHz that was also available in the United Kingdom was divided into 12 blocks of 10 MHz. Each bidder was allowed to bid for two or three blocks. A bidder that was the high bidder on one block at the end would not win, nor would it owe anything, and the unsold block would be immediately re-auctioned. The regulator did not report all bids, but only the identity of standing high bids at the end of each round.

Seven bidders applied to participate in the German auction: the four incumbents—T-Mobile, Mannesmann (Vodafone), Viag (O2), and E-Plus—as well as three potential entrants. The auction did not reserve any blocks for entrants, and indeed was criticized beforehand for allowing the incumbents to foreclose entrants. The auction thus allowed for four, five, or six winners.

One of the entrants, Debitel, dropped out when the price reached DM[15] 4,897 million per block in round 121. This was not common knowledge for all the bidders, as only high bids, and not all bids, were reported by the regulator after each round. However, bidders could guess after a few rounds that Debitel had dropped out. Once the remaining six bidders had concluded that one bidder had dropped out, these six bidders could have settled right away for 20 MHz, or, instead, keep bidding, hoping to win 30 MHz and face less competition in the market after the auction.

Mannesmann and T-Mobile elected to keep going for another 53 rounds, until prices reached DM 8,330 million. At that time T-Mobile found it more prudent to drop out, and risk ending with two blocks and a six-player market, than keep going. Had Mannesmann won three blocks, there would have been one unsold block, which T-Mobile could have tried to acquire in a second auction.[16]

The results of the German auction are in table 9.4. Each of the six winning bidders won two blocks.

This was not the real end of the process. Neither of the entrants that *won* spectrum, Mobilcomm (backed by France Telecom) and Group3G (back by Telefonica) ever elected to roll out service. Both abandoned their licenses, and for no refund or other considerations. Germany was left with a four-player market.[17]

The Italian auction (table 9.5) used the UK design; it apparently also achieved a reasonably efficient outcome despite the very limited competition—there were six bidders for five licenses.

The Austrian auction copied the German design, but there were six, rather than seven, bidders to start. That auction settled in seven rounds. Apparently on one wanted to repeat the German experience, and consolidation occurred after the auction.

Table 9.5
Italian 3G auction

3rd	23rd				
day	October 2000				
11th round	Start: 9,30	Finish: 9,50			
		Bid			
Bid order	Bidder	Lire$_{\times 1.000}$	Euro	Remaining waivers	Expression of interest*
1	OMNITEL	4.740.000.000	2.448.005.702	3	
2	IPSE	4.730.000.000	2.442.841.133	3	Yes
3	WIND	4.700.000.000	2.427.347.426	3	
4	ANDALA	4.700.000.000	2.427.347.426	3	Yes
5	TIM	4.680.000.000	2.417.018.288	3	
6	BLU	4.490.000.000	2.318.891.477	2	

The SMR and simultaneous auction format has now been used in dozens of auctions for spectrum—almost all the spectrum auctions in both North and South America, Australia, and parts of Asia, including India, Pakistan, Singapore, and Taiwan. It has also been used for a number of energy auctions; perhaps the first successful application was the simultaneous descending clock auction (SDCA) used for energy procurement in New Jersey,[18] and then Italy, Illinois, Montana, Ohio, and elsewhere, and for various other energy products in Austria, France, and Germany.[19]

9.5 SMR Auction Configuration

Experience with the SMR auction format has not been uniformly successful. What follows provides some examples of outcomes that appear inefficient—in some cases larger and more valuable lots sold for less than smaller lots. This section explains why at times different bidders pay different prices for identical aggregate amounts, and why bidders winning more can even pay less. In general, the auction prices and allocations appear to leave some bidders overpaying relative to others and suggest that misallocations or inefficient allocations occurred.

What follows are several examples of how auction configuration (and in particular activity rules are implemented), and in combination with bidding strategy, can affect the outcome of an auction.

9.5.1 The Texas Capacity Auctions

Perhaps the most glaring example of what can go wrong is seen in a few auctions for generation capacity in Texas in 2001 (table 9.6 and 2002 (table 9.7).[20] The first capacity auction included one-year and two-year contracts.

These auctions were for 25 MW blocks of capacity for specific durations. The capacity was owned by three companies: CPL, Reliant, and TXU. These generation companies were under a regulatory mandate

Table 9.6
Texas capacity auctions: fall 2001

	Reliant South	TXU South	Δ	%Δ
One-year base load	$7.32	$10.59	$3.27	45%
Two-year base load	$6.59	$10.33	$3.74	57%
One-year base cyclic	$0.75	$1.79	$0.95	127%

Table 9.7
Texas capacity auctions: spring 2002

	Reliant South	CPL South	Δ	%Δ
June base load—capacity only	$13.13	$6.85		41.3%
June base load—capacity + fuel	$22.57	$18.76	$3.81	
July base load—capacity only	$6.76	$10.60		45.3%
July base load—capacity + fuel	$22.57	$18.76	$3.06	
August base load—capacity only	$16.76	$9.85		55.6%
August base load—capacity + fuel	$26.86	$22.79	$4.07	

to sell off a fraction of their capacity, which they did in a series of quarterly auctions starting in 2001. The tables here are for the first two auctions.

In the fall 2001 auction, the Reliant and TXU capacity was for *identical* blocks from the same physical, jointly owned, base load generators. These blocks were sold in an SMR auction. In that auction, Reliant and TXU separately managed the starting prices and bid increments, and assigned separate eligibility points for each "auction." Reliant started with a lower price, but TXU set lower increments. After several rounds of bidding, the TXU prices passed the Reliant prices. As a result the Reliant part of the auction cleared first. When the TXU price subsequently passed the Reliant price, bidders that had stopped bidding on the Reliant product were stuck. There was no provision in the auction rules to allow bidders to switch back and forth.

The second capacity auction was for monthly contracts. A similar pattern arose in that subsequent spring 2002 auction. Again, the CPL and Reliant products were identical. The bids were for capacity, but fuel costs were added in separately. Bidders failed to adapt in the second Texas capacity auction. Eventually, the Texas Public Utility Commission mandated that a switching rule be introduced.[21]

9.5.2 The Spanish 4G Auction (2011)

The 2011 Spanish 4G auction provides another striking example of what problems can arise with the activity requirements in SMR auctions that include both regional and national licenses.

This auction (table 9.8) included nine bands within the 2.6 GHz band. Seven were for national licenses (four each of 2 × 10 MHz, and three each of 2 × 5 MHz), and the other two were divided into 19 regions (one license of 2 × 10 MHz and one of 2 × 5 MHz in each region).

Table 9.8
Spanish 4G auction (2011)

Coverage	Winner	Price of 2 × 10 MHz	Price of 2 × 5 MHz	Totals
National	Orange	€ 22,214,958		€ 45,096,364
National	Orange	€ 22,881,406		
National	Telefonica	€ 21,791,816		€ 44,438,132
National	Telefonica	€ 22,646,316		
National	Vodafone		€ 9,811,904	
National	Vodafone		€ 9,789,654	
National	Vodafone		€ 10,150,457	€ 59,095,246
Regional	Vodafone		€ 29,343,231	

Table 9.9
Spanish 4G auction: three large regions, 5 vs 10 MHz prices

Region	Population	Price of a 2 × 10 MHz license	Price of a 2 × 5 MHz license
Cataluña	7,512,381	€4,239,078.48	€5,886,062.90
Madrid	6,545,684	€3,818,047.06	€5,043,280.86
Pais Vasco	2,178,338	€2,417,683.79	€3,186,291.36

The three largest bidders each won 2 × 20 MHz. However, there were many ways for this outcome to emerge. What happened was that Vodafone won a combination of national 2 × 5 MHz and regional 2 × 5 MHz licenses, whereas the two other incumbents, Orange and Telefonica, each won national licenses. The main decision facing these operators in this auction was when to switch from the national to the regional licenses. The activity requirements made it almost impossible to switch back. Until one switched to regionals, there would also be little competition to cause prices to increase. So the decision to switch had to based on an estimate of how hard the regional operators would compete to get the 2 × 5 MHz blocks in each region. The outcome suggests that Vodafone, which ended up switching first, guessed incorrectly, as it paid almost 30 percent more than its rivals (see table 9.9).

What is also striking in the Spanish auction is that in each of the three most expensive regions, Catalunya (including Barcelona), Madrid, and Pais Vasco, the price of the 2 × 5 MHz license was more than the price of the 2 × 10 MHz license.

Table 9.10
FCC auction 66: final prices paid by six largest winners

	MPOPs	POPs	PWB total	Ave $/MPOP
Cricket + Denali	2,201,624,760	176,183,964	$985,082,750	$0.447
Spectrumco	5,287,189,470	267,387,437	$2,337,609,000	$0.451
Cingular	2,436,458,880	198,768,198	$1,334,610,000	$0.558
T-Mobie	6,638,718,070	474,718,308	$4,182,312,000	$0.630
Cellco (VZW)	3,840,952,220	192,047,611	$2,808,599,000	$0.731
MetroPCS	1,445,444,020	144,544,402	$1,391,410,000	$0.963

This was also a result of the limitations imposed by a combination of the activity rules and spectrum caps. The activity rules did not allow a bidder to easily arbitrage between the prices of the national and regional licenses. And there were spectrum caps that did not permit a bidder to bid for 2 × 10 MHz when 2 × 5 MHz was more expensive.

9.5.3 The Mexican PCS Auction

Another quite striking example of how activity rules can limit arbitrage across difference size licenses, in this case of 10 MHz and 30 MHz licenses in the Mexican 2G auction (table 9.11). The activity rule specified that the 30 MHz licenses counted for five times as many points as the 10 MHz licenses. This made switching difficult in one more way: a bidder dropping from 30 MHz to 10 MHz might not be able to switch back. The next effect was that in one region, 7, the 30 MHz licenses sold for more than the 10 MHz licenses.

Activity points that had permitted bidders to more easily arbitrage the price differentials between the 10 MHz and 30 MHz licenses would have prevented this type of anomaly from occurring. It is not clear that there was a benefit in assigning more points to the 30 MHz blocks than the 10 MHz ones in any region. The goals of the activity rules include facilitating efficient arbitrage and also compelling bidders to mark increasingly serious offers. Assigning a larger number of points to the larger regions probably accomplishes this goal.

9.5.4 US FCC Auction 66—Advanced Wireless Services

The US AWS auction 66 was another auction in which activity rules limited arbitrage (table 9.10).

In that auction the FCC allocated six bands of spectrum partitioned geographically in three ways: one 20 MHz block was allocated in each

Table 9.11
Mexico 2G auction results

Block (MHz)					
Region	Population	A(30)	B(30)	D(10)	E(10)
1 – Baja	26,216,350	$161,544,000	$162,523,000	$152,400,000	$137,870,000
2 – Sonora	4,375,593	$29,560,000	$31,642,000	$15,000,000	$14,870,000
3 – Chihuahua	4,978,464	$101,656,000	–	$70,100,000	$70,000,000
4 – Monterrey	7,493,285	$616,877,000	$588,488,000	$363,300,000	$353,252,000
5 – Campeche	8,256,749	$9,000,000	–	$6,800,000	$6,700,000
6 – Guadalajara	10,934,908	$289,699,000	$286,811,000	$89,650,000	$81,077,000
7 – Agua Caliente/ Guanajato	10,339,977	$70,876,000	$75,941,000	$82,800,000	$82,121,000
8 – Oaxaca	18,382,181	$20,397,000	$20,160,000	$6,300,000	$6,400,000
9 – DF	23,742,926	$892,500,000	$941,506,000	$485,100,000	$479,160,000

of 734 cellular market areas (CMAs), one 20 MHz and one 10 MHz in each of 176 economic areas (EAs), and one 20 MHz and two 10 MHz licenses in each of 12 regional economic area groups (REAGs). The average prices were highest for the REAG licenses[22]; they were over $0.10 (20 percent) more than for the CMA and EA licenses. Also the average price by bidder varied quite a bit. Auction 66 had such a range of licenses that assigning points to prevent activity traps very difficult. To address this issue some recent (combinatorial clock) auctions (CCAs), a *revealed preference* activity rule has been imposed. A reveal-preference rule requires bids in any one round be consistent with those in previous rounds.[23]

While each bidder won a different footprint, some comparisons can be made. T-Mobile won 20 MHz licenses in approximately 80 percent percent percent of the top 100 markets. In contrast, Spectrumco covered essentially all of the top metro areas and paid approximately 25 percent percent less than T-Mobile. Cellco and Cingular combined to cover the entire United States; Cricket and MetroPCS also covered all the major metropolitan areas.[24] Clearly, as in the Spanish 4G auction, the choice of whether to purchase the larger REAGs or the smaller EAs or CMAs affected the average price. In contrast with Spain, the prices in this auction for the large regions were higher.

The outcome of auction 66, while perhaps difficult to predict, can be explained at least partly by the bidding strategies. The two parties achieving the lower average prices (the Cricket-Denali consortia and Spectrumco) were apparently better able to determine when to switch from the larger REAGs down to the smaller—$0.451 for Spectrumco and $0.630 for T-Mobile. There were two approaches used to make this determination.[25] Spectrumco employed jump bids to try to assess the budgets of at least marginal bidders. In many auctions the total exposure—that is, the sum of the values of all bids, both high bids and bids that have been topped—can peak long before the auction closes. This peak can be used to provide an estimate of the budgets of all bidders in the auction.

If one bidder *knows* its rivals' budgets, then it can use this information to achieve a better allocation and lower prices.[26] Consider the following example:

Suppose that there are two bidders and three lots. Each bidder has values for each lot of 100 (or more) but is quite budget constrained, having a total budget of 100. If one bidder can get its rival to bid 50 or more for one lot, it can win the other two lots. And the more it can get

Table 9.12
Two bidder and three lots: two budget-constrained bidders and three lots

Bidder/value for	Lot 1	Lot 2	Lot 3	Budget
Bidder A	100	100	100	100
Bidder B	100	100	100	100

that rival to sink in the one lot, the lower will be the price it has to pay for the other two. The multi-round structure of an SMR auction allows for this type of strategy. It is the type deployed by Spectrumco in Auction 66.

This budget-based strategy has some risks. It can be difficult to forecast budgets. And even the ability to forecast budgets is no guarantee of retaining the flexibility to take advantage of it. Moreover it is unusual for all bidders to face effective budget constraints.

Another approach in Auction 66 and in the Spanish auction is to try to forecast the conditions under which the small blocks will sell for less than the larger blocks. This is the approach taken by Cricket and Denali. If there are both large bidders seeking larger regional licenses and small bidders seeking only a few licenses, then the large bidders will want to stay on the larger blocks until they get to be more expensive than the sum of the smaller blocks. At some critical premium the large-block bidders will want to shift to regional blocks, anticipating that they will end up being a better value. Wether this should occur will depend on the relative amounts of competition for the larger blocks and the smaller blocks. To some extent this can be observed during an SMR auction.

9.5.5 Arbitrage in SMR Auctions

The preceding examples indicate that arbitrage opportunities can be limited in SMR auctions. These examples raise two immediate questions. One is whether there is harm—that is, whether there is a loss of revenue for the auctioneer, or bidders are harmed. The other is what measures can be taken to prevent any such harm.

Proposition 2 *Consider an SMR auction with at least two bidders and two lots in which activity rules do not allow bidders to switch between substitutes, at least in some stages of the auction. Then, for some preference profiles, this auction will necessarily result in inefficient outcomes even with straightforward bidding.*

Proof Suppose there are at least two lots, or combinations of lots, and at least two bidders, and that one bidder that was topped on one block is no longer eligible to switch to the other. If the auction were to end, then there would be an alternative allocation in which this bidder would get the other block and the bidder with the second block would not. This alternative allocation would result in higher bidder payoffs and revenues whenever the bidder that should not win has a lower value for that second block than the bidder that should win. ∎

Recall Milgrom's result that the SMR auction with straightforward bidding will result in an efficient outcome. This results requires certain conditions about activity requirements to be met. What the proposition above means if that the activity rules do not permit arbitrage, then the outcome can be inefficient.

Now there are many ways in which the activity rules can limit arbitrage possibilities. And other auction design and management decisions can have a significant impact on the outcome of an auction. What follows is a summary of the impact of these decisions.

1. *Switching rules* The Texas capacity auctions did not allow any switching. Once Texas introduced switching rules, the price discrepancies disappeared.
2. *Abstract blocks* Other auctions, such as the Spanish and German 4G auctions and the German 3G auctions, assign separate labels and auction prices to various identical abstract blocks. Clearly, combining offers for all identical abstract blocks (assuming an SMR auction or having a single price for all the blocks in a simultaneous clock auction) can only improve efficiency.
3. *Starting prices, bid increments, and activity parameters* In simultaneous clock auctions, relative prices are determined by the auctioneer. The starting prices and bid increments determine how fast the prices of different lots increase. What often happens is that the larger lots tend to clear first. This tends to turn a simultaneous auction into a sequential one, as bidders' decisions are essentially sequential.[27]

For example, in FCC auction 66 there were three types of geographic licenses with 20 MHz and covering New York City: a 20 MHz Northeast REAG had a bit less than 48,000,000 bidding units, the 20 MHz New York EA had just over 24,000,000 bidding units, and the 20 MHz New York CMA had approximately 16,000,000 bidding units. An activity requirement of 60 percent would permit switching from the New York CMA to the New York EA license, but one of above two-thirds would not. And this would also allow switching from New York and either the Boston or the Philadelphia EA to the Northeast REAG covering those EAs. Thus the choice of activity requirements can facilitate or hinder arbitrage.

4. *Activity points* As the preceding discussion makes clear, the way activity points are set, in combination with the activity requirement percentage, can affect auction efficiency. Generally, when a number of small blocks combine to make one large block, then the small blocks should have a combined number of activity points equal to that of the larger one. Such a provision allows a bidder to switch from a sent of small blocks to a large block with same bandwidth and geographic coverage. And this facilitates arbitrage. The above suggests that the way points are assigned to the small blocks, in combination with the activity percentage requirements, will affect auction efficiency.

5. *Size of licenses and blocks* When there are both regional and national licenses in the same auction, it will not usually be the case that the sum of the regional prices will equal the price of a national license. Similarly, when there are smaller and larger blocks, such as 10 and 20 MHz licenses, it will not usually be the case that the price of the smaller blocks sum to exactly the price of the larger one. Any activity rule, except for one allowing package bids or one based purely on revealed preferences, will limit arbitrage possibilities,

6. *Pace of the auction* An auction can be too fast or too slow. Spectrum auctions can take several months to complete. The temptation for regulators is to speed up bidding. However, the analysis and time it takes a bidder to make a decision suggests that speeding up an auction can result in bidder error, reduced revenues, and total welfare. Nevertheless, energy procurement auctions, which span more than one day, tend to be too slow. The bidders must hedge positions and holding open such positions over night can be very costly.[28] Also some auctions grant bidders a limited number of *round extensions* to extend time for submitting bids in a round, whereas other provide a limited number of *activity waivers* for a round. The difference is the extension extends the round for all bidders, whereas the waiver just affects a single bidder, but does not affect the round schedule. With a large number of bidders and lots, round extensions would tend to slow an auction more than would waivers.[29]

9.6 Conclusion

This chapter has described perhaps one of the most significant contributions economic theory has made to regulation and policy—the simultaneous multiple auction. This auction format has the potential to offer significant improvements in auction revenues, and for bidders, by facilitating better matches of bidders and lots. However, there are a number of ways in which an SMR or simultaneous clock auction can be misapplied and result in an undesirable outcome. This chapter has provided descriptions of the main SMR auction pitfalls and some suggestions for avoiding them. The SMR auction is best suited for auctions of substitutes and some types of complements. Package bidding, which allows bidders to avoid exposure problems, is described in chapter 10.

10 Combinatorial Auctions

Summary

This chapter looks at package bidding, or combinatorial, auctions. Bidders in multi-object auctions will at times be bidding for substitutes and/or complements. This gives rise to significant additional complexity for both bidders and the auctioneer. When bidders are considering substitute licenses or packages, they will want to obtain the best package for the best price. This need not happen, and both bidders and the auctioneer can suffer. When there are complements, a bidder seeking a large package may end up facing an exposure problem, that is, it may win part of the desired package and pay more than it is worth. This problem can deter bidders from bidding close to value. And when a small bidder is competing against a package bidder, the auction rules can tilt the outcome toward the package bidder or the small bidder.

10.1 Introduction

This chapter looks at package bidding, or *combinatorial*, auctions. Bidders in multi-object auctions will at times be bidding for substitutes and/or complements. This gives rise to significant additional complexity for both bidders and the auctioneer. As seen in the previous chapters, the SMR and clock auctions are very well suited for achieving efficient outcomes when all bidders view all objects as substitutes. However, these auctions can still leave bidders facing some difficult decisions when there are complements, at least for some lots and some bidders in the auction.

When bidders are considering substitute licenses or packages, they will want to obtain the best package for the best price. This need not happen, and both bidders and the auctioneer can suffer. When there are complements, a bidder seeking a large package may be deterred from bidding close to value to avoid paying more for just a part of the package than it is worth. And when a small bidder is competing against a package bidder, the auction rules may tilt the outcome either toward the package bidder or the small bidder.

To better illustrate the issue, suppose there are two lots, and that there is one large bidder interested in both lots and several small bidders each interested in only one of the lots. And suppose these two lots are being sold in an SMR or a simultaneous clock auction.

Let L denote the large bidder's value for the two lots, let s^{j1} denote the highest value among the small bidders for lot $j = 1, 2$, and let s^{j2} denote the second highest value. Suppose that the large bidder has zero value for one lot.

If $L > s^{11} + s^{21}$, then the large bidder should always win. But this may not happen if, for example, the difference between L and $s^{11} + s^{21}$ is not very large or if the large bidder is very reluctant to risk winning only one lot. This is a simple example of the *exposure* problem. Absent package bids, the large bidder has to take a risk to have any chance at winning.

A more complex case arises with three bidders and three lots (table 10.1). Bidder A only wants the package of lots 1 and 2, B only wants the package of lots 2 and 3, and C only wants the combination of lots 3 and 1.

In the first, two-unit example, if the auctioneer were to call out prices for lots 1 and 2, p^1 and p^2, for which $L \geq p^1 + p^2$ and for which $p^j > s^{j1}$ for $j = 1, 2$, then the large bidder would be willing to purchase both lots, and the small bidders would not be willing to purchase any lots. In this case these prices would be a *competitive equilibrium* in that, when all bidders act as price takers, demand exactly equals supply.

In the second example, with three lots and three bidders, there is no competitive equilibrium. To see this, suppose the contrary. Let p^j,

Table 10.1
Three-bidder, three-lot auction

Bidder/value for	Lots 1 and 2	Lots 2 and 3	Lots 1 and 3
Bidder A	1	0	0
Bidder B	0	1	0
Bidder C	0	0	1

$j = 1, 2, 3$, be such an equilibrium. Then the following conditions must hold:

$$p^j \geq 0 \tag{10.1}$$

for all $j = 1, 2, 3$. This says that prices cannot be negative. Also

$$p^j + p^k \leq 1 \tag{10.2}$$

for $j \neq k$, $j, k = 1, 2, 3$. This says that no bidder will pay more than value. For there not to be any excess demand, the expression (10.2) has to hold with equality for all bidders. But this cannot hold for any set of prices.

Package bidding allows bidders to place all-or-nothing bids on a set of lots. Thus, in the example above, bidders can place a single all-or-nothing bid on a pair of lots. Bidder A can place a bid of 1 for the package of lots 1 and 2, B can bid 1 for 2 and 3, and C can bid 1 for 3 and 1. This is an equilibrium, with one bidder randomly winning its bid.

However, package bidding is not a perfect solution. First, if there are small bidders whose values add up to more than that of a large bidder, then these small bidders might need to find a way to coordinate their bids. This is the *threshold* problem. To be more specific, consider the case in which $L < s^{11} + s^{21}$, and there are separate bids for lot 1, lot 2, and the package of both lots. If the stand-alone "winning" bids for lots 1 and 2 together add up to less than L, then the small bidders would each need to raise their "winning" bids to top L. However, each small bidder would want the other to raise its bid more, and absorb a larger share of the gap. This free-riding problem might prevent the small bidders from winning even when they have the higher valuations.

Package bidding also is inherently more complex than restricting auctions to those in which there is a separate price for each lot, or product, in the auction. If there are n lots available, there will be 2^{n-1} available packages. Other than for very small numbers of lots, or when the types of complementarities are well understood and limited, package bidding can quickly become computationally infeasible. The process for expressing preferences (how can a bidder communicate values for all such packages?) and then finding the winning set of offers can be unmanageable.

This does not mean that package bidding is never helpful. What follows provides some specific types of package auctions, some of which have been used in practice.

10.2 Hierarchical Package Bidding

Perhaps one of the simplest forms of package bidding is one in which bids are allowed on individual lots and on a few predefined packages. This has been called *hierarchical package bidding* (HPB).[1] The US Federal Communications Commission (FCC) conducted one HPB auction, auction 73,[2] that raised over $18.9 billion. In auction 73, the C block band was divided into twelve regional economic area group (REAG) licenses. Bidders could bid on individual REAG licenses or the package of all twelve REAGs.

The HPB auction was otherwise virtually the same as an SMR auction. Licenses were assigned points. Bidders had to establish initial eligibility by placing up-front deposits for their maximum desired number of points prior to the auction. The HPB auction imposed similar activity requirements: high bids plus new bids had to exceed a given fraction of the eligibility, or the bidder lost eligibility in proportion to the gap between actual activity and the activity target. Prices increased from one round to the next for all licenses for which there was excess demand. And the activity requirement was increased in stages as prices rose and aggregate eligibility dropped during the course of the auction. The closing rule was also the same: the auction would only close on all licenses at the same time and only when there was no excess demand for any license or package.

There were a few differences. First, a bidder could place a bid on a license and on a package that contained that license. However, such a bidder would not get credited with double the activity. Second, if there was more bidding activity on the packages than on the component licenses, it was possible that the provisional winning bid amounts after a round for a package exceeded the sum of the provisional winning bids for the individual licenses in the package. Therefore some rule was needed to determine minimum required bids for the component licenses, especially where there was no excess demand for some of those components. And bidders seeking packages were not permitted to bid the component prices if those were lower. The FCC developed a formula for allocating the gap in determining minimum required bids for the component licenses.[3] Finally, a bidder who had stopped bidding on one of the component licenses might find that, after several rounds of bidding in which other components in the package prices increased, their old previously losing bid could turn into part of a provisional winning bid. That bidder might have lost eligibility, or used that eligibility for other licenses. The SMR rules were modified to allow that bidder to resume bidding on that component but not otherwise increase eligibility.

Goeree and Holt (2010) conducted laboratory experiments, with Caltech students acting as bidders, comparing the SMR, HPB, and unconstrained package auctions. In their experiments they found that the HPB allowed bidders in some cases to solve the exposure problem, and therefore achieved greater efficiency and revenues than the SMR auction. Note that the efficiency measure was based on the difference between the actual sum of the values of the final assignment and the maximum possible sum. They also found that unconstrained package bidding offered no additional benefits.

These experiments *assumed* a particular structure of complementarities. They did not allow for arbitrary or random complementarities. Nor did they identify when the threshold problem might become more significant than the exposure problem. What these experiments do suggest is that where there is a well-defined exposure problem—where different bidders have similar views of the complementarities—then the HPB seems to be an effective and efficient mechanism. In the actual auction the package bidder won the national license.

Goeree and Holt's (2010) experiments were conducted on behalf of the FCC in advance of FCC auction 73. The FCC had previously conducted experiments on a different form of package bidding, called the *simultaneous ascending auction with package bidding* (SAAPB).[4] The results of those experiments showed improvements over the SMR auction similar to those with the HPB. However, the SAAPB did take significantly longer to run than the SMR. Perhaps because SMR auctions can already be quite long (commonly several weeks, but sometimes over four months), the FCC elected to use the HPB over the SAAPB. No tests were run to compare how many rounds an SAAPB takes on average compared to the HPB.

10.3 VCG Auction Properties

The multi-object version of the VCG auction introduced in chapter 4 is one type of package auction. As was explained in chapter 4, the VCG has the property that bidders will generally have strong incentives to report true values. Thus the outcome of a VCG auction should be efficient. However, as is explained in this section, the VCG auction has a number of other features that could block its adoption.[5]

This section also provides a few examples illustrating these undesirable properties. Note that bids in a VCG auction need to convey to the auctioneer the bidders' valuations for all possible combinations of lots. Thus an individual bid is a set of reported values for each package, or combination of lots, that is available in the auction. In general, this list can be quite large—with n lots, the number of packages is $2^n - 1$. So the VCG auction may be somewhat impractical if the number of lots is large. But, as the following examples illustrate, the VCG has other drawbacks even where there are few lots for sale (table 10.2).

10.3.1 Revenues in VCG Auctions

One difficulty with VCG auctions is that they may fail to generate much revenue. To see this, recall that bids in a VCG auction are all-or-nothing offers for packages. The winning bids are those mutually consistent bids that maximize the sum of the offers, and not necessarily the auction revenues. The reason is that the price a bidder pays is the difference between the sum of the rivals' reported offers for what they would have won without this bidder, and their reported offers for what they actually do win. The following examples illustrate how revenues are determined, and in some cases, can be zero.

First, consider the following example (table 10.2) in which there are two objects and three bidders. In this example it is assumed that bidders can place bids for one and/or both lots.

Bidders B1 and B2 each want only one of the two lots available. They are each willing to pay 100 for it. Bidder B3 wants both lots or none, and will pay up to 50 for the two lots. As both B1 and B2 individually block B3—that is, B3 loses even if either B1 or B2 does not bid—then the VCG prices for B1 and B2 are each zero. So low revenue can be a feature of VCG auctions.

Simple variations of this example indicate that the VCG has other not so appealing features. First, consider the case where values are now lower for B1 and B2 (table 10.3). B3 wins both lots. So, if B1 and B2 were really weaker than B3, but both overbid (or colluded to do so), then they could win both lots for much less than their value.

Table 10.2
Three-bidder, two-lot VCG auction

Bidder/bid for	1 lot	2 lots
Bidder B1	100	100
Bidder B2	100	100
Bidder B3	0	50

Table 10.3
Three-bidder, two-lot VCG auction

Bidder/bid for	1 lot	2 lots
Bidder B1	20	20
Bidder B2	20	20
Bidder B3	0	50

Table 10.4
Two-bidder, three-lot VCG auction

Bidder/bid for	1 lot	2 lots	3 lots
Bidder B1	100	200	280
Bidder B2	100	120	130

Day and Milgrom (2008) explain how the outcome is sensitive to mergers and demergers. If, for example, bidders B1 and B2 were to merge and be treated as one bidder that wants both lots, then the merged entity would win both lots at a price of 50 rather than 0. So a merger works to increase the amount that a bidder may have to pay, and a demerger can reduce it. Indeed, rather than splitting into two entities, a combined B1-B2 bidder can enter a second shill bidder. By doing so, B1-B2 can reduce the price it has to pay. This also raises the question whether the auctioneer would want to stop the shill from bidding. In most auctions, more bidders result in higher prices paid. But this need not be the case with VCG auctions. Competition can lower price.

10.3.2 Fairness in VCG Auctions

The VCG auction may result in efficient but unfair outcomes. Arguably, one bidder might be troubled if it wins less than a rival bidder but pays more. The following examples show that this can happen in theory and has happened in practice.

Consider the case of two bidders and three lots (table 10.4). The Vickrey price a bidder pays will be the difference between the value its rivals get at the final allocation and what the rivals would have won had that bidder never submitted a bid. In table 10.4 there are three lots for sale, and two bidders. Each bidder submits bids for one, two, or three lots. Bidder B1 wins two lots and B2 wins one. The value of the

Table 10.5
Swiss 2012 auction

Frequency band	Orange	Sunrise	Swisscom
800 MHz	20 MHz	20 MHz	20 MHz
900 MHz	10 MHz	30 MHz	30 MHz
1,800 MHz	50 MHz	40 MHz	60 MHz
2.1 GHz	40 MHz	20 MHz	60 MHz
2.1 GHz TDD	0 MHz	0 MHz	0 MHz
2.6 GHz	40 MHz	50 MHz	40 MHz
2.6 GHz TDD	0 MHz	0 MHz	45 MHz
Amount paid (CHF)	154'702'000	481'720'000	359'846'000

two lots to B1 is 200. B1 also indicated that all three lots together were worth 280. B2 wins one lot which it values as being worth 100, but would have paid 120 for two lots and 130 for all three.

Thus B1's VCG price is 30, as B2 only offered an additional 30 for the two lots that B1 wins. And B2's VCG price is 80. This is clearly a bit odd, and B2 may view it as unfair. VCG auctions will not generally have the property that a bidder winning more than a rival pays less. Indeed the reverse can occur. Notice too that that the outcome is efficient in that it maximizes the sum of the bidder values across all feasible allocations.

Table 10.5 shows that this unfairness is not just a theoretical possibility. It shows the outcome of a Swiss spectrum auction.

The first four categories (800 MHz, 900 MHz, 1,800 MHz, and 2,100 MHz) were the most valuable, and the 800 MHz and 900 MHz were more valuable than the 1,800 MHz or the 2,100 MHz. Each block was 10 MHz, and there were six in the 800 MHz, seven in the 900 MHz, the equivalent of fifteen in the 1,800 MHz, and twelve in the 2,100 MHz category.[6] While this was not a one-shot sealed-bid auction, in that it permitted multiple bids in multiple rounds plus a separate set of sealed bids (as will be explained in more detail below), it used the VCG pricing rule, with a reservation price. As can be seen in the example, Swisscom won more blocks and paid more than 30 percent less than Sunrise. Orange paid the reservation price.

10.3.3 Budgets in VCG Auctions

The VCG auction is particularly poorly suited to cases where bidders face budget constraints. The following example has three bidders and three lots, and assumes budgets are binding for all bidders:

Three-bidder, three-lot VCG auction			
Bidder/bid for	1 lot	2 lots	3 lots
Bidder B1	100	100	100
Bidder B2	100	100	100
Bidder B3	100	100	100

In this example each bidder wins one lot and pays zero. Now, if one bidder has entered a bid of 101 for two or three lots and zero for one lot, then that bidder could win two lots for 100. This auction has one type of unstable pure strategy equilibrium. This would have one bidder offering 200 (or more) for all three blocks, and the other bidders offering zero each. Of course, the other bidders could offer more than zero, and the bidder making the large offer would then want to revise its offer. But there will be no set of offers that constitute a stable (perfect) Nash equilibrium.[7]

To see this, suppose that there is a set of bids, one for each bidder, which constitute a Nash equilibrium in this example. Each bidder will want to allocate its budget among the three blocks. If each of two bidders allocates its budget of 100 to one block, then, as above, a rival can win two lots by offering 101 for two lots, and zero for other combinations. If only one bidder allocates its budget to one lot, then the other two bidders will be competing for the remaining two lots. And the only possibility is that they each allocate their budgets to one lot. A similar analysis would establish that there is no way for the three bidders to each allocate their budgets to the three lots so that no one bidder would not want to revise its allocation.[8]

The experience in many spectrum auctions indicates that even in very large auctions with large publicly traded firms, budgets can be constraining.[9] So this example can be quite relevant.

10.4 Core-Selecting Auctions

This section addresses some of the issues raised above. First, it describes conditions under which the final prices are not so low that a losing bidder would be willing to offer more than the winners have to pay. When an auction has this property, then the auction outcome is said to be in the *core*.[10] More generally, the auction outcome is in the core when there is no group of bidders that can bid together and exclude other bidders, and that together would then each have a better outcome *that the auctioneer would also prefer*, that is, one that would result in higher auction revenues.

Notice in the examples above that when an auction results in a non-core allocation, at least one bidder viewed the lots as complements. Notice, in particular, that the value of a second lot is higher than that of the first. Bidders B1 and B2 each individually *blocked* B3. So, when B1's and B2's bids were each high enough to block B3, then B1 and B2 will each have a zero Vickrey price. If the lots were also substitutes for B3, so that the value of the second lot was no higher than the first, the outcome would be in the core. B1 and B2 would each pay the value of the first lot to B3. Given that the value of the first lot for B3 was more than the value of the second, what B1 and B2 pay in total (twice B3's value of the first lot) is necessarily more than B3's value of both lots. So B3 no longer blocks the outcome, and the allocation is in the core.

More generally, it has been shown[11] that when all goods are mutual substitutes, the auction outcome will be in the core, and then the Vickrey auction outcome is in the core.[12] As the preceding discussion and examples demonstrate, this need not be the case when the goods are complements. The reason is that when one bidder's offers for a set of lots is larger than the sum of its offers for the individual lots, or for nonintersecting subsets, then two or more rivals can individually prevent that bidder from winning. So, if each of those rivals offers enough to block the bidder with complements, then their Vickrey prices will not reflect the value of the losing bid. This type of situation cannot arise when the lots are always substitutes for all bidders.

One additional property of core-selecting auctions is that the outcome does not change if bidders merge, or if a bidding consortium breaks up. The logic is fairly straightforward. Suppose that some set of bidders could earn more bidding as one bidder than if they split up into separate bidders. More specifically, let S_1 and S_2 be disjoint subsets of the set of all bidders, N, and suppose that S_2 consists only of shill bidders, so that $w(S_2) = 0$, where $w(S)$ denotes the amount that the coalition $S = [S_1 \cup S_2] \subset N$ can guaranty for its members when bidding together in the auction. Suppose that $w(S) > w(S_1)$. Note that efficiency of core allocations implies that $w(T) + w(S) = w(N)$ and $w(S_1) + w(S_2) + w(T) = w(N)$, where $T = N \setminus S$. This implies that $w(S_2) = 0$; that is, the set S_2 of shills gets nothing on their own. Now consider the merging of S_1 and S_2 into S. Then $w(S) + w(T) = w(N)$. But also $w(S_1) + w(S_2) + w(T) = w(N)$.

So $w(S_1) + w(S_2) = w(S)$. Clearly, mergers do not increase profits. Summarizing:

Proposition 3 (Milgrom 2004) *An efficient direct auction mechanism has the property that no bidder can ever earn more than its Vickrey payoff by disaggregating and bidding with shills if and only if it is a core-selecting auction mechanism.*

10.5 Ascending Package Auctions and the Combinatorial Clock Auction

The *combinatorial clock auction* (CCA) is a form of ascending package auction that specifically addresses the low revenue of VCG auctions, and it can lead to a core outcome. An ascending package auction is a simultaneous auction in which prices rise from round to round, as in an SMR auction, but all bids are package bids. When bidding is straightforward, as when using surplus-maximizing proxy bidders, Ausubel and Milgrom (2002) have shown that the final allocation of an ascending package auction is in the core. This section describes the CCA.

The Swiss auction described in the previous section was a CCA. The CCA has also been used for selling spectrum licenses in a number of other countries, including Austria, Denmark, Ireland, and the United Kingdom, and most recently in Australia and Canada.

The CCA differs from an SMR auction in a number of important ways:

- Before a CCA (as before an SMR auction), the available lots are assigned points, and bidders must specify how many points of initial eligibility they want. Applying for more points usually requires larger financial guarantees, which can provide some disincentive for applying the maximum number of points.
- In each round a bidder has to submit new bids that meet its activity target. In other words, the activity must meet the activity target. In most CCAs, a bidder's activity target is 100 percent of its eligibility. If a bidder's bids fall short of this target, its eligibility is reduced in all future rounds to the new, lower activity.[13]
- All bids are package bids. This is unlike an SMR auction, in which some lots bid in a round can become provisional winning bids. Indeed, it is common in SMR auctions that final winning bids for two or more licenses are made in different rounds.
- Any bid from any round can become a winning bid.

- The clock phase ends, as in SMR auctions, when there is no longer any excess demand for any product.
- Unlike in an SMR auction, the clock phase of a CCA does not end the auction. Bidders can submit bids in a "supplementary round" of bidding. The bids in the supplementary round must improve, in some sense, on bids made during the clock rounds, and be consistent.[14]
- In a CCA, all bids are mutually exclusive. While a bidder can submit different package bids that do not overlap, a bidder can only win one bid. So, if a bidder wants to submit an offer for two different packages A and B, it might want to submit separate offers for A, B, and A combined with B.
- At the end of the supplementary round, winners are determined by maximizing the sum of bidder offers across all feasible combinations of bids made during the clock rounds and during the supplementary round. The prices paid are the VCG prices except when these prices are not in the core—in which case some rule is applied to adjust prices to ensure that they are in the core.[15]

I have summarized the main elements of the CCA rules. CCAs have now been used in perhaps a dozen or more auctions. The rules differ across auctions, especially in regard to the core adjustment and to how activity and consistency requirements are implemented in the supplementary round. What is generally required is that the offers in the supplementary round satisfy a version of a *revealed preference* rule.

What follows is an example illustrating a revealed preference activity rule. Suppose, for example, that a bidder was bidding for x units of product A and y units of product B. Further suppose that during the clock phase this bidder always bids on two units of A and two units of B when the price of A was no more than Δ_1 larger than price of B, and never bids on more than one unit of A when the price differential reached $\Delta_2 > \Delta_1$, that is,

$$D(p_A, p_B) = \begin{cases} (2,2) & \text{if } p_A - p_B \leq \Delta_1, \\ (1,3) \text{ or } (2,2) & \text{if } \Delta_1 < p_A - p_B < \Delta_2, \\ (1,3) & \text{if } p_A - p_B \geq \Delta_2. \end{cases}$$

This bidding pattern indicates that the bidder has indicated that it would prefer the second lot of product A over a third lot of B when the price differential is less than Δ_1, and would prefer the reverse when the price differential exceeds Δ_2. Precisely where this bidder would switch is not necessarily revealed during the auction.

A revealed preference rule would require that supplementary round offers for the package (2, 2) never exceed the offer for (1, 3) by more than Δ_2 or less than Δ_1. The implementation of the revealed preference rule, often called a *relative cap rule* or RCR, requires a number of decisions on how to resolve potential inconsistencies during the clock rounds, as is discussed below.

What follows is a brief review of some properties of the CCA, and a few examples from actual auctions.

10.5.1 Complexity

The CCA allows bidders to submit bids on any or all packages of lots available in the auction. The number of combinations can be astronomically large—if there are N objects for sale, then there are $2^N - 1$ possible combinations. Finding winners and determining prices requires choosing, from among all mutually consistent sets of offers from all the bidders (i.e., offers for which total demand does not exceed supply), the set that maximizes the sum of the reported values. Further, solving for prices requires recalculating the optimal solution when each bidder is excluded. If the are n^j offers from bidder j, $j = 1, 2, \ldots, J$, then there are $n^1 \times n^j \times \ldots \times n^J$ possible combinations of bids to compare. Unless these numbers are very small, bidders cannot possibly submit bids for all combinations, nor can the auctioneer calculate winners and prices.

This complexity problem is of some practical concern. For example, in the Swiss auction described above, there were ten categories of licenses (across seven bands). There were over 60 million possible packages for each of the three bidders, and over 2×10^{23} possible combinations in all.[16] Auctioneers are forced to limit the number of combinations, and bidders cannot fully express preferences. So, as a practical matter, the CCA is not a useful mechanism for eliciting information from bidders about preferences.

Nor is the CCA the most efficient way of eliciting information about values. For example, suppose that there are four products, and a bidder wants at least 1 unit of each product and a total of 6 units. The bidder may also want to purchase up to 4 units of each product. The CCA would require the bidder to submit values for 256 combinations. In practice, the bidder may have a value for the base package of one unit of each product, and then an incremental value for each additional unit. If that is the case, the bidder would really need to submit only 13 values.

10.5.2 Bidder Incentives

While the CCA is complex, it does provide bidders with some incentives to report true values—in that it uses Vickrey pricing. What follows examines where bidder incentives may differ from those in a standard Vickrey auction.[17] The main differences between a VCG auction and a CCA arises when bidders payoffs depend not just on the price they pay but also on what rivals pay, and when there is a core adjustment.

Fairness In the Swiss auction, Sunrise won fewer blocks and paid much more than Swisscom. Reportedly, Sunrise complained to the regulator. Unfortunately for Sunrise, the Swiss regulator, Bakom, had been made aware of this possibility and had indicated it was not relevant for its decision to proceed with the CCA.[18] But a firm's decision makers and their advisers are evaluated based on relative performance. Thus inequitable outcomes may be economically inefficient when bidders' preferences include relative performance.

The first two large multi-product CCA auctions both resulted in some bidders paying more than rivals that won more spectrum. The discrepancy in the Swiss auction was quite large. There was also a smaller but quite direct discrepancy in the Dutch 4G auction (table 10.6). In that auction KPN won the same number of abstract blocks in the 800 MHz, 900 MHz, and 1,800 MHz bands as did Vodafone. However, KPN also won 25 MHz of additional spectrum in the 2,600 MHz band. Yet KPN's base price was €30 million less than Vodafone's. To avoid relatively poor performance in a CCA, bidders may need to deviate from bidding based purely on economic surplus as measured by their business models.

Bidding Long In a CCA, as in Vickrey auctions, what rivals offer affects the amount any given bidder will pay. The CCA differs from a Vickrey auction in that there are multiple bidding rounds, and bidders are provided information about rivals between the rounds. The relative cap rule means that at the end of the clock phase a bidder may not have to offer full value to ensure what it wants in the auction. By not bidding full value, a bidder can improve its relative price. However, a bidder can be forced by a rival to bid more for what it wants, when the rival keeps the clock phase from closing. The following example (table 10.7) assumes three bidders seeking four lots.[19]

Table 10.6
Dutch 4G auction results

Band/bidder	KPN	Vodafone	T-Mobile	Tele 2
800 MHz	2 × 10 MHz	2 × 10 MHz		2 × 10 MHz
900 MHz	2 × 10 MHz	2 × 10 MHz	2 × 15 MHz	
1,800 MHz	2 × 20 MH	2 × 20 MH	2 × 30 MHz	
2,100 MHz	2 × 5 MHz	2 × 5 MHz		
1,900 MHz			14.6 MHz	
2,600 MHz TDD	25 MHz		25 MHz	
Base price	€1,349,851,000	€1,380,793,000	€910,582,000	€160,813,000
Assignment round delta	€2,001,000	€7,000	€99,000	
Total price paid	€1,351,852,000	€1,380,800,000	€910,681,000	€160,813,000

Table 10.7
Bid long

Round	Price/ bidder	Bidder 1 demand	Bidder 2 demand	Bidder 3 demand	B1 bid amount	B2 bid amount	B3 bid amount
1	45	4	2	2	180	90	90
2	50	4	2	2	200	100	100
3	55	4	1	1	220	55	55
4	60	2	1	1	120	60	60
8	80	2	1	1	160	80	80

If the auction were to end after round 4, then, assuming no supplementary-round bids, bidder B1 would win two lots, and B2 and B3 would each win one. B1 would pay 80, because B2 and B2 had, in round 3, offered a combined 200 for the four lots but only won one lot each, worth a combined 120. However, if the clock phase were to end four rounds later at a price of 80, then B1 would only pay 40. Note that B2 and B3 can offer up to 110 each in the supplementary round for two lots (as the maximum allowed offer is based on the price at which the bidder reduced eligibility). So B1 may not necessarily benefit so much from extending the clock rounds, and will risk winning all the blocks. But there are clearly incentives to depart from straightforward bidding.

Budget Constraints When bidders have fixed budgets, a CCA and a VCG present very challenging decision problems for the bidders. The first example is a variation of the chopsticks problem analyzed by Szentes and Rosenthal (2003).[20] Suppose that there are two bidders and three lots. Further suppose that each bidder has a budget of 1, and has to enter a separate bid for each of lots 1, 2, and 3.[21] Suppose too that the amount a bidder pays is the amount the rival bidder would have paid for its lots. Further suppose that the value of each lot exceeds a bidder's budget. If a bidder's total offer ever exceeds its budget, the Vickrey price can also exceed its budget. So, if a bidder is not allowed to take any such risks, the bids for one or any combination of the three lots can never exceed 1 in total. But this means that each bidder is virtually certain to pay less than 1, and will have money left over. There will never be a pure strategy equilibrium in this game—the only possible equilibria involve randomization.

Now consider the same bidders in a CCA. The bidders will tend to bid up to their budgets and then drop. If bidders always behave in this manner, their winning offers and their maximum offers will equal their budgets. Then the Vickrey prices are always zero.[22] In experiments and in practice, it does appear that bidders do have flexible budgets or take some risks. The data in most actual auctions are confidential, and so it cannot be verified to what extent bidders risk exceeding budgets. Experience suggests, however, that such behavior does occur.

Experimental data suggest that bidders will take very limited risks.[23] In a CCA, bidding above budget can be nearly riskless, as prior-round bids make the probability of those risky bids' winning essentially zero. Nevertheless, it does appear that bidders cannot easily assess risks, and they bid cautiously. The net effect is that payments will tend to be below budgets.

10.6 Comparing the Performance of the CCA and Other Auctions

There are now a large number of papers comparing the performance of the different auction formats.[24] What follows is a basic analysis of the differences. To compare a CCA and an SMR auction, suppose that there are ten lots available, one large bidder that will purchase all ten lots, and two small bidders each seeking one lot (table 10.8).

If this auction were to be conducted as an SMR or clock auction, in which bidders paid their bid amounts, bidder B1 would win but must pay a round 4 price of 60 for each lot, for a total of 600. In contrast, in a CCA this bidder would have to pay only 200, or at most 220, for all ten lots. Thus the CCA appears to provide an advantage to the large

Table 10.8
Comparing CCA and SMR auctions

Round	Price/ bidder	Bidder 1 demand	Bidder 2 demand	Bidder 3 demand	B1 bid amount	B2 bid amount	B3 bid amount
1	45	10	2	2	450	90	90
2	50	10	2	2	500	100	100
3	55	10	1	1	550	55	55
4	60	10	0	0	600	60	60

bidder. And if a large bidder wants to foreclose the small one, the costs of doing so may be much less in the CCA than in the SMR auction.

This is not just a theoretical example. One of the first CCA auctions was one for the UK L band (table 10.9) in May 2008.[25] That auction included 17 lots: 16 small ones, and one large one, which was roughly equivalent to three small lots.

The last clock-round prices were £871,000 per lot, or more than £16,500,000 in aggregate. However, QUALCOMM, which won all the lots, paid £8,334,000 for all 17 lots. QUALCOMM's final-round bid was even higher, £20,000,000. Revenue equivalence does not hold in this case. Revenue equivalence does not tend to hold when bidders are bidding for multiple lots. To date, the theoretical, empirical, and experimental evidence on CCA vs SMR auction revenues is inconclusive.

10.7 The Austrian CCA

This section provides a brief review of a recent combinatorial clock auction in Austria. The Austrian CCA was especially interesting as it appears both spectrum caps and budgets may have played a significant role in the outcome. The Austrian CCA was very similar to the Dutch one (figure 10.2) in that there were three incumbents bidding and almost the entire prime spectrum was included in the auction. The main differences from the Dutch CCA were that no entrants showed up in Austria, so that there were six blocks available for incumbents in the 800 MHz band rather than four in the Netherlands and there were spectrum caps in Austria.

The Austrian auction included a total of 28 blocks, all 2 × 5 MHz—seven blocks in the 900 MHz band and 15 in the 1,800 MHz band in addition to the six 800 MHz blocks. Bidders could win, and bid for, up to four blocks of 800 MHz, six blocks of 900 MHz, a combined seven

Table 10.9
UK L band auction

Bidder	LA	LB	LC	LD	LE	LF	LG	LH	LI	LJ	LK	LL	LM	LN	LO	LP	LQ	Usage	Bid	Opportunity cost	Base price
Qualcomm	1	1	1	1	1	1	1	1	1	1	1	1	1	1	1	1	1	H	20,000	8,334	8,334
Bidder	Without Qualcomm																	Usage	Bid		
ePortal								1	1	1	1	1	1	1				L	3,382		
Vectone			1	1	1													H	2,338		
WorldSpace																	1	H	2,614		
ePortal								1	1	1	1	1	1	1				L	3,382		
Vectone			1	1	1													H	2,338		
WorldSpace																	1	L	2,614		

blocks of 800 and 900 MHz spectrum, and no more than 14 of the available 28 blocks. As there were three incumbents bidding, this auction appeared to be designed, possibly not by intent, to be a game of knockout.

Standard approaches to valuing a mobile franchise suggest that the value of winning 14 blocks can exceed 50 percent more than the value of winning 9 or 10 blocks when winning 14 blocks means one rival is likely to win no spectrum, or very little, reducing post-auction competition (see ITU Report 2012). This is also consistent with standard models of oligopoly theory (Tirole 1988). Thus, if a bidder in the CCA would want to place true values for 14 blocks and a post-auction two-player market and also true values for 9 or 10 blocks with three players in the market post-auction, then it would be likely that there would be only two winners.

Most of the actual bid data are confidential. So it is impossible to discern from publicly available data whether the three competing mobile network operators, Telekom Austria, T-Mobile, and Hutchison, bids reflected a premium for knocking out a rival. However, the Austrian regulator did provide some information about bidding patterns in the Supplementary Round, as reported by Georg Serentschy (2013) head of the Telecom Control Commission (TKK) after the auction.

The clock phase ended with a significant amount of overshooting, namely unallocated blocks. More specifically, 14 percent of the blocks received no bids in the last clock round. The three bidders submitted a total of 4,032 bids In the Supplementary Round. Serentschy (2013) states that more than 69 percent or approximately 2,600 of these Supplementary Round bids were on the 14 block packages.[26] The 14 block packages were for half the available spectrum, and presumably intended to knock out a rival. While there were also bids for other numbers of blocks, it does appear the three bidders concentrated their

Table 10.10
Bidders in French 4G menu auction

Band	Bouyges	Free	Orange	SFR	Total amount paid (€M)
2,600 MHz	€228	€272	€287	€150	€937
Blocks won	2 × 15 MH	2 × 20 MH	2 × 20 MH	2 × 15 MHz	
800 MHz	€683		€891	€1,065	€2,639
Blocks won	2 × 10 MHz	nono	2 × 10 MHz	2 × 10 MHz	

Table 10.11
Outcome of the 2013 Austrian CCA

Bidder	Telekom Austria	T-Mobile	H3G	Auction total
800 MHz blocks	4	2	0	6
900 MHz blocks	3	3	1	7
1,800 MHz blocks	7	4	4	15
MHz won	140 MHz	90 MHz	50 MHz	280 MHz
Final price	€ 1,029,895,738	€ 653,998,071	€ 329,031,581	€ 2,012,925,390
Reserve price	€ 289,600,000	€ 182,400,000	€ 72,900,000	€ 544,900,000
Euros/MPop	€ 0.87	€ 0.86	€ 0.78	€ 0.85

bids on 14 block knockout bids. Further, according to the Serentschy, the price offers for the knock out bid packages were largely at the limits allowed by auction rules, whereas the other bids were bid at 60 to 70 percent of the ceiling prices.

10.8 Conclusion

This chapter examined some forms of package bidding. Package bidding is potentially a significant practical concern for bidders. VCG auctions, a form of package bidding, have some unfortunate properties, though they do provide bidders incentives to bid truthfully, and they achieve efficient outcomes. These properties have appeared in actual auctions.

The problem of low revenues can be addressed through core selection auctions. However, predefined packages may be more appropriate in certain cases, as has been seen in some of the experimental literature.

One other type of package bidding that has been proposed is menu auctions.[27] In a menu auction bidders are asked to report values for different packages. For example, if bidders can win any number k of blocks with $\underline{k} \leq k \leq \bar{k}$, then a bid is an amount offered for each such possible k. This type of auction provides bidders incentives to bid true incremental values—even in a pay-as-bid (i.e., first-price) auction.

The reason that menu auctions have this property is that the bidding level determines the overall discount that a bidder applies, and then each bidder will have an incentive to only bid incremental values for incremental blocks. However, menu auctions may have multiple equilibria.[28] Menu auctions also present a faster and simpler sealed-bid approach to package bidding.[29]

A version of the menu auction was used in the French 4G auctions (table 10.10). There were two auctions, one for the 800 MHz bands and one for the 2,600 MHz bands. In each case bidders were asked to indicate demand curves for a range of 2 to 6 blocks, and for particular blocks in case they were not equivalent. The winning bids were based on the set of offers that maximized the sum of the total values (with some weights used to reflect other objectives of the auctioneer, e.g., coverage). Bidders were required to pay bid amounts.

11 Final Remarks

Summary

This chapter provides an overview of the main points covered in the preceding chapters. The intent is to provide some guidelines to what aspects of auction theory and experience may be relevant in any particular situation.

11.1 Introduction

This book is intended as a Primer and not an encyclopedic guide or a survey of auction theory and practice. The intent is to provide some guidelines to what aspects of auction theory and experience may be relevant in any such situation. This chapter highlights some of the key issues for the various types of auctions encountered in practice.

11.2 Game Theory and Bidder Incentives

Any analysis needs to start with an assessment of the incentives and constraints of the bidders. This is true whether the analysis is intended to guide either a bidder or the auctioneer perspective. Forgetting to take account of incentives will result in surprises for both the bidder and the auctioneer. It is all too common for an auction design to fail in this way. The standard online English auction with a fixed closing rule is a classic example in that it creates large incentives for bidders to wait unit the last second. This makes the final outcome quite random.

In multi-attribute auctions, scoring rules can create incentives to distort values, leading to very perverse outcomes. This is what happened in the California QF auction, in which the suppliers were paying to supply energy, even during high-demand hours. In multi-object auctions the structure of the auction can limit arbitrage opportunities or even present bidders with incentives to distort demand. There are many examples where this has occurred—the New Zealand spectrum auctions in the 1980s were one. But even the Google ad auctions for position limit bidders' ability to arbitrage price differentials for substitutes. This was also the case in the Mexican 2G and Spanish 4G auctions, in which activity rule parameters forced bidders to make sequential decisions; the result was that in some regions, larger and more valuable licenses sold for less than smaller licenses.

Budget constraints in multi-object auctions can confront bidders with insoluble decision problems. For example, in a multi-object, sealed-bid auction, whether pay-as-bid or Vickrey, a bidder will be required to predict final prices in determining how to bid. This is part of what happened in the New Zealand auction for radio stations, cited above. But the auction design can be modified to make bidding decisions much simpler. For example, bidders can be allowed to submit demand schedules rather than separate bids for each object.

11.3 Bidder Incentives and Revenue Equivalence

Perhaps the most fundamental insight of auction theory is the revenue equivalence theorem (RET). The form of the pricing rule affects bidder incentives—at times, in such a way that two seemingly quite different auctions can result in the same outcome. The debate about whether to use a pay-as-bid or a second-price auction often fails to take account of the RET.

This is not to say that one can always rely on the RET to say that auction design and bidding strategy do not matter. The RET applies only in some situations, and it is more limited in its applicability to multi-object auctions than to single-object auctions. Moreover, even when the RET applies, the decision problem facing a bidder need not be the same in two different auctions that should result in the same outcome. For example, in a second-price auction for a single object, bidding value is a dominant strategy. The decision problem facing a bidder is relatively straightforward. However, even in the case of independent private values, where the RET does apply, a bidder in a first-price sealed-bid auction has to solve a complex optimization problem to determine its optimal bidding strategy, unless the bidder knows the RET and can easily calculated the expected value of the second highest value bidder conditional on winning. So, while theory would predict outcome equivalence, bidder decisions are more straightforward, and as a result it would appear that the outcome should be more predictable in a second-price auction than in a first-price auction.

11.4 Single-Object versus Multi-Object Auctions

Bidders face inherently more complex decisions in multi-object auctions than in single-object auctions. When there are substitutes, bidders will need to decide what objects to bid more aggressively for. When there are complements, bidders face exposure problems. The form of the auction, including the nature of the bid, the pricing rule, and other provisions, has an enormous effect on the decision problem facing bidders, and therefore the auction outcome. Auctioneers want to carefully consider options for the auction format and its influence on bidding behavior as well.

One of the most significant successes in putting economic theory into practice has been the development of the SMR auction and other simultaneous auctions that simplify decision problems facing bidders. Simplifying decision problems can result in more efficient outcomes, higher revenues for the auctioneer, *and* more surplus for the bidders.

Package bidding, or combinatorial auctions, can be utilized for situations in which bidders face complementarities. In those cases in which there are known complementarities that are the same for many bidders, and little conflict over which packages are best, using predefined packages can be desirable in an SMR auction. However, this requires that the auctioneer have some knowledge about bidder preferences, and that is not always the case. The Vickrey pricing rule appears particularly unattractive for combinatorial auctions, especially when bidders have budget constraints. Package bidding can also introduce significant complexity into the bidding process. The complexity itself can result in more bidding errors and adversely affect the outcome.

11.5 Sequential Auctions

Auctions are not especially one-time events: similar or identical objects are often auctioned at different times. This means that bidders in an early auction will want to forecast prices in deciding how to bid. One very central result of auction theory, the martingale theorem, states that the expected price in one auction is the realized price in the previous auction. However, when bidders' opportunities and values can change over time, this result needs modification. In that case the decision how to bid can matter. And the decision about how much to auction at each date can affect the outcome for the auctioneer.

Sequential auctions can occur for complements too. The early winner will tend to have an advantage in subsequent auctions. This is especially true in auctions for renewing franchises where there are complementary sunk and durable investments. Prices in the first auctions will tend to be above values, in subsequent auctions below values; they will not provide much direct information about true underlying values.

11.6 Bidding Behavior

Most of the analysis in this book has assumed, explicitly or implicitly, that bidders act rationally to maximize their expected payoffs. This assumption is clearly unrealistic in many situations, even in very large auctions with very sophisticated bidders. However, starting with this benchmark is also useful.

Perhaps most useful is working through the decision problems facing bidders. Referring again to the comparison between first-price and second-price auctions, it is clear that bidding strategy is generally simpler in a second-price auction—at least for single-object auctions. This is not to say that behavior does not matter. In experimental settings, bidders at times can systematically depart from rational play.[1] At other times, the outcome may not be so sensitive to deviations.[2] Where there is a lot at stake, bidders have time to analyze the decisions, and the decision problems are straightforward, then the assumption of rational bidding is likely to be most valid. However, it should also be clear from this Primer that an analysis of the decision problem facing a rational bidder can also be useful even where bidders lack the ability, resources, incentives, or time to come up with optimal bidding strategies. Indeed at times the costs and effort required for coming up with an optimal decision can far outweigh the potential benefits.

When a bidder faces very random outcomes, then the decisions are especially difficult. Some auctions, those that rely on outguessing rivals, have intrinsic randomness, in that they only have mixed equilibria. It is often possible to design an auction to avoid this randomness. This type of uncertainty can, in most cases, only result in worse expected outcomes for both the bidders and the auctioneer.

11.7 Optimal Auctions

A fundamental contribution to auction theory is Myerson's (1981) paper on optimal auction design. Auctions inherently involve some informational asymmetries: bidders know their own values, which are unknown to the auctioneer and possibly to each other. While bidders may end up revealing values in any optimal auction, this does not mean that when asked to report values, bidders will necessarily do so. Moreover, even in Vickrey auctions, in which bidders have strong incentives to report true values, the auctioneer will not necessarily maximize its expected revenues. Indeed the optimal auction from a seller's perspective will necessarily include a reservation price.

The optimal auction—for example, one that maximizes seller revenues—will necessarily leave some surplus for the bidder. This is due to the underlying informational asymmetries—asymmetries that leave information rents to the bidders. Thus the goals of eliciting truthful valuations from bidders and maximizing auction proceeds can conflict, even if bidders always end up bidding in a way that can be used to infer true values.

11.8 Auction Design, Management, and Strategy

Putting the principles described in this Primer to use involves many decisions. These decisions can be very complex. The goal in this Primer is not to provide a checklist to allow a bidder or an auctioneer to decide on the best bidding strategy or best auction format. Rather, this Primer is intended to provide the bidder or the auctioneer with a way of starting an analysis of its decision problem. The basic tools of game theory, in chapter 2, should be applied. And the analysis of auction design or strategy will depend on whether there is a single object or multiple objects (which can be substitutes or complements) and on whether each bid will include a single lot or multiple lots. However, good design and management decisions and good bidding strategies will almost always require some analysis based on the tools given, though no formula can be applied in all situations. And even where formulas can be applied, care is needed to avoid misuse.

11.9 Participation

As a final remark, auctions don't work without bidders. Not only must there be bidders, the bidders need to submit competitive and meaningful bids. Auction rules and management can make bidder decisions difficult, which discourages bidders from making meaningful bids. This is true in even simple auctions, such as simultaneous sealed bids. But it is also true in more complex auctions, such as combinatorial clock auctions, bidders can be confronted with impossibly difficult decisions. Auctions can fail if bidders fail to be informed of the opportunity, or if bidder qualification criteria or payment rules are too lax.[3]

11.10 Auctions and Markets

This Primer is intended to illustrate some pitfalls but, more important, to explain how bidder incentives need to be factored into any analysis of auction design and bid strategy. This Primer explains how the auction design can affect bidder incentives, which in turn affect the auction outcome.

Auctions are a form of market mechanism. Auctions differ from other types of markets in that they have formal rules governing offers, allocations, and pricing. An auction is particularly well suited for one-off or intermittent transactions with relatively few bidders. Higher frequency transactions typically occur in other types of markets and exchanges.

That auction design, management, and strategy can have a large impact on the final allocation and prices suggests that auctions are neither reliable nor predictable. This Primer should make it clear, however, that auctions, if managed well, can be a useful and efficient type of market mechanism for pricing and making allocation decisions.

Notes

Chapter 1

1. See Paarsch and Hong (2006).

2. See Krishna (2009) and Milgrom (2004) for more theoretically advanced treatments and also Klemperer (2004).

3. Sometimes an auction originator will engage an independent third party to serve as an auction manager. However, the auction manager usually assumes an operational role. In contrast, the exchange specialist, or trading desk, will serve to adjust prices to balance supply and demand offers, and its operations are generally housed by an exchange that handles hundreds or thousands of types of assets.

4. See Wolfstetter (1996).

5. See Garber (1989) and Van den Berg, Van Ours, and Pradhan (2001).

6. The SMR and SAA auctions and the related SDCA allow for simultaneous auctioning in a sequence of rounds of multiple lots. Bidders topped in one round can respond in the next. Details are explained in chapter 9.

7. This includes the empty set and the set itself.

8. See Mueller (1993) and Salant (2000).

9. See chapter 4 and Milgrom (2004) for a discussion.

10. See Joskow and Kahn (2001) for a discussion, and Gribik (1995).

11. See http://energycommerce.house.gov/comdem/legviews/mvbrspec.htm.

12. Klemperer (2004, p. 191).

13. See Capen, Clapp, and Campbell (1971) for a discussion of oil lease bids with common values.

14. See Salant (2005) for a discussion.

15. See Milgrom (2004).

16. Cramton and Schwartz (2000) document one case discussed below, and Brusco and Lopomo (2002) explain the theoretical risks of open bidding. Also see Marshall and Marx (2007).

17. See Klemperer (2004) or Cramton and Schwartz (2000) for a discussion.

Chapter 2

1. This concept of dominant strategies is discussed below.

2. To see that there may be no pure strategy equilibrium, suppose that each bidder has the same value v, and that each bidder j offers a_j, b_j, $(v - a_j, b_j)$ on blocks $j = 1, 2, 3$. If $a_j \geq b_j \geq 1 - a_j - b_j$, then the other bidder can bid a bit more than j on the second and third blocks and win both.

3. See Larson and Salant (2003) and Hortacsu and Puller (2008).

4. What follows is a very simple example. Suppose that two suppliers in a region each have capacity of 800 MW. This means that both firms are needed, and either firm can always supply 400 MW at the ceiling price of 100. Normalize price so that demand net of operating costs (assumed to be the same for both) is 1,200 MW for $p \leq 100$ and zero otherwise, where p denotes the price. In a symmetric equilibrium, each firm will choose a probability distribution function $F(p)$, which makes the rival indifferent as to what price it chooses. Neither firm will ever offer a price above 100—or below 50, as for prices below 50 a firm can earn more by offering the ceiling price. Thus $400pF(p) + 800p[1 - F(p)] = 4{,}000$, or $F(p) = 2p - (100/P)$, for $50 \leq p \leq 100$. For $p < 50$, we have $F(p) = 0$, and for $p > 100$, $F(p) = 1$. Each firm choosing its price with a probability determined by $F(p)$ is an equilibrium in this example.

5. See http://www.arcep.fr/fileadmin/reprise/dossiers/4G/proj-dec-appel-800mhz-160511.pdf and http://www.arcep.fr/fileadmin/reprise/dossiers/4G/proj-dec-appel-2-6ghz-160511.pdf.

6. The details of this process are described in Gribik (1995).

7. The circle around the column player nodes in figure 2.1 indicate that the column player does not know whether the row player has chosen Top or Bottom when it is choosing between Left and Right.

8. A further requirement that can be imposed on equilibrium in a multi-stage game is that the equilibrium be *stable* for various forms of perturbations. Small variations in payoffs can at times disrupt some SPEs. Further small changes in strategies can also do the same. The ideas of a *trembling hand* perfect equilibrium (see Selten 1975) and the *stable set* of equilibria (see Kohlberg and Mertens 1986 provide further refinements of the SPE criterion).

9. See Brusco and Lopomo (2002) and Cramton and Schwartz (2000).

10. See Brusco and Lopomo (2002), Cramton and Schwartz (2000), and www.fcc.gov/wtb/auctions.

11. Signaling games, such as those in which bidders send signals in auctions, typically have a multiplicity, and often a continuum, of equilibria. See Van Damme (1991) for a discussion of signaling games.

12. See, for example, Joskow and Kahn (2001) for a description of one such market.

13. See Friedman (1985) for a discussion of cooperative equilibria in finitely repeated games.

14. See Abreu, Pearce, and Stacchetti (1990).

15. For a more detailed discussion of cooperative equilibria with demand or cost uncertainty, see Green and Porter (1984).

Chapter 3

1. This form of auction is sometimes referred to as a Japanese auction, or a Japanese-style English auction.

2. This type of auction is sometimes referred to as a Yankee auction.

3. See Vickrey (1961) and Riley and Samuelson (1981).

4. How quickly a bidder should bid can also be part of the strategy, as it can influence other bidders.

5. See Capen, Clapp, and Campbell (1971).

6. The following discussion assumes independent private values. If there are affiliated values, it is still a dominant strategy for a bidder to bid its value, but this value is conditional on it being the high bidder.

7. The first-order conditions are that the optimal b will satisfy $(v-b)P'(b)-P(b) = 0$, or $(v-b)/b = P(b)/bP'(b)$.

8. See Kahn et al. (2001).

9. It is assumed that bidders use equilibrium bid strategies and have the same preferences toward risk.

10. Table 3.1 shows the results of the bidding in the largest ten regions.

Chapter 4

1. To see this, notice that both bidders will have valuations above $\frac{1}{2}$ with probability $\frac{1}{4}$. In this case expected revenues are. They both have valuations below $\frac{1}{2}$ with probability $\frac{1}{4}$, resulting in the object being unsold. And then, with probability $\frac{1}{2}$, only one bidder can meet the reservation price, and revenues equal the reservation price. This results in expected revenues of $\frac{1}{4}\times\frac{2}{3}+\frac{1}{2}\times\frac{1}{2}+\frac{1}{4}\times 0=\frac{5}{12}$.

2. In chapter 10 properties of VCG auctions are explained in further detail.

3. See Green and Laffont (1977) for a more formal discussion.

4. Also see Rothkopf et al. (1990) for an extensive discussion. Rothkopf and colleagues indicate that original Vickrey second-price auction is not always incentive compatible when there is more than one object for sale, or when bidders are asymmetric. The $\frac{2}{3}$ generalized VCG auctions described in this chapter remain incentive compatible in these cases as well.

5. See Mueller (1993).

6. The following is from Milgrom (2004).

7. See Levin and Skrzypacz (2013) for further discussion.

8. See http://stakeholders.ofcom.org.uk/spectrum/spectrum-awards/ for examples of such auctions in the United Kingdom.

9. See http://en.itst.dk/spectrum-equipment/Auctions-and-calls-for-tenders/3g-hoved mappe/3g-auction-2001-1.

10. There was no document released by the Danish government indicating the reason for this provision. It may be the case that this auction design could result in

higher prices, or at least, it would be less obvious when a bidder paid much less than its value.

11. See Mueller (1993). It is unclear from Mueller whether the auctions were simultaneous or sequential.

12. Varian (2007) gave a figure of over $10 billion for 2005, just for Google and Yahoo.

13. The pivot mechanism has each bidder pay the difference between the aggregate value of rivals' winning bids without that bidder and the value with that bidder.

14. Much of what follows is based on Milgrom (1996) and joint work done on a related project.

15. See http://transition.fcc.gov/wcb/tapd/universal_service/JointBoard/welcome.html.

16. See Milgrom (1996) for details.

Chapter 5

1. See Capen, Clapp, and Campbell (1971) or Hendricks and Porter (1988) for a discussion of oil lease auctions.

2. In contrast, in a pure private values case, each bidder's valuation is independent, meaning there is no common value term, v.

3. Suppose, for example, that forecasts are uniformly distributed on some interval $\left[V-\frac{1}{2}, V+\frac{1}{2}\right]$, where V is the unknown true value. Then the expected value of V for a bidder that knows it has the lowest forecast value s^N among N bidders is $s^N - \frac{1}{N+1} + \frac{1}{2}$.

4. If there are only two bidders, then this analysis needs to be modified, as in this case the second lowest bidder would want to wait longer.

5. If there are only two bidders, or if the number of lots available is more than half the number of bidders, then the marginal losing bidder will tend to have a forecast value that is below the true value.

6. See Klemperer (2004).

7. See, for example, Hendricks and Porter (1988).

8. See Milgrom and Weber (1982).

9. Suppose that the strong bidder places a value that is 2 higher than the uninformed bidder's, and that the value to the informed bidder can be 2 or 8. If the probability of a high value is sufficiently great, namely $2p - 4(1-p) > 2(1-p)$, where p is the probability of a high value, then the strong bidder will always win, and will bid 8. This bidder will win 2 with probability p and lose 4 with probability $1 - p$. For smaller values of p, the strong bidder will prefer to allow the informed bidder to win when the value is high.

10. See Porter (1983).

11. A few years later, Telefonica reentered the German market through an acquisition of one of the *winners*, and was losing money as of 2013.

12. See Milgrom and Weber (1982).

Chapter 6

1. See, for example, www.bgs-auction.com. These auctions stagger purchases over time. Each auction also contains a provision for deferring procurement to a later date should participation in any one auction be limited. See also Loxley and Salant (2004).

2. See, for example, www.globaldairytrade.info for wholesale dairy auctions, and www.cranberryauction.info for wholesale cranberry auctions.

3. Other studies of sequential auctions include Ashenfelter and Graddy (2004), Deltas and Kosmopoulou (2004), Jones, Menezes, and Vella (2004), Lyk-Jensen and Chanel (2007), and Mezzetti (2011).

4. See Ashenfelter (1989).

5. See Ashenfelter (1989) and Ashenfelter and Graddy (2003).

6. See Milgrom (2004, p. 95).

7. See Loxley and Salant (2004) for a more extensive discussion.

8. In particular, the firm participating in the forward market is a *Stackelberg leader* (see Tirole 1988).

9. This example follows Bernhardt and Scoones (1994).

10. If this were a third-price auction, that is, the highest two bids were to win and pay the third highest bid amount, then each bidder would have a strong incentive to bid its value. The RET states that the expected outcome would be the same as with a pay-as-bid, sealed-bid auction.

11. This is the pricing rule equivalent to a second-price rule for an auction with a single object.

12. This issue was addressed by McAfee and Vincent (1993).

13. This is assuming additively separable utility.

Chapter 7

1. The US Federal Communications Commission has been awarding most of the broadband PCS licenses for ten-year terms. See the license-period provisions at www.fcc.gov/wtb/auctions.

2. See www.caiso.com for a discussion of the California capacity auction proposal. See www.iso-ne.com/markets for a description of the ISO New England capacity auctions.

3. If $v_1 > v_N + K$, then bidder 1 could win the second auction after losing the first auction; however, in equilibrium, bidder 1 would win the first auction.

4. This assumes that $2v_2 > v_1$.

5. Much of this follows Riordan and Salant (1994). See also Vickers (1986) and Gilbert and Newbery (1982).

6. A similar argument implies that the lead firm would still win if it had costs c^2 and the other firm had costs c^1.

7. Here drastic means an innovation so large as to make the old technology economically irrelevant in the market.

8. See Vickers (1986) or Gilbert and Newbery (1982).

9. Also see Jehiel and Moldovanu (2003) for more detailed discussion.

Chapter 8

1. In a reverse auction the price sequence would be decreasing and the offers would be supplies at each price.

2. See Walras and Jaffé (1954).

3. Some of what follows is in Milgrom (2004).

4. When bidders have affiliated values, this value is condition on being a high-value bidder.

5. See Cramton and Schwartz (2000) and Grimm, Riedel, and Wolfstetter (2003).

6. See Hortacsu and Puller (2008), Wolak (2003), and Borenstein, Bushnell, and Wolak (2002).

7. See Green and Newbery (1992), Baldick, Grant, and Kahn (2004), and Larson and Salant (2003).

8. See Baldick, Grant, and Kahn (2004) for a discussion.

9. This section is loosely based on Klemperer and Meyer (1989).

10. If marginal costs are positive, the first-order condition becomes $[D(p) - S^i(p) + (p - C'((D(p) - S^i(p))][D'(p) - S^{i\prime}(p)] = 0$.

11. As long as demand is concave and costs are convex increasing, the second-order conditions will also be satisfied.

12. See Larson and Salant (2003).

13. See McAdams (2007).

Chapter 9

1. See http://wireless.fcc.gov/auctions/ and Milgrom (2004). Note that for spectrum auctions, the term *allocate* is usually reserved for the type of use of the spectrum (television, mobile communications, public safety, etc.), and the term *assignment* is used for the process whereby individuals or firms are awarded rights to use the spectrum.

2. See Kwerel and Rosston (2000).

3. As discussed in more detail below, concerns about signaling and collusive bidding led the FCC and other regulatory agencies to adopt limited-disclosure bidding. See Cramton and Schwartz (2000) and Brusco and Lopomo (2002). Also Marshall and Marx (2007) discuss bidder collusion in SMR auctions.

4. The Indian 3G auction in 2010 was an exception.

5. This section assumes a forward auction. In a reverse auction, prices start high and then fall.

6. See Loxley and Salant (2004) for a description of a simultaneous descending clock (reverse) auction.

7. See Cramton and Schwartz (2000) and Brusco and Lopomo (2002) for a discussion.

8. Fisher (1922) is an early reference for the concept of sufficient statistics.

9. See Cramton and Schwartz (2000). Also see Fan, Cuihong, and Wolfstetter (2013) for a different analysis of the impact of information disclosure provisions.

10. See McAdams (2007) for a more formal analysis for the case in which the auctioneer does not have to sell both, or all, lots.

11. See Loxley and Salant (2004).

12. The FCC's second spectrum auction was not an SMR auction.

13. A bid withdrawal in the South region resulted in one of the blocks selling for a significant discount with respect to the others. However, if one adds back the withdrawal penalties that needed to be paid, the total price for this block was very similar to that for the others.

14. The 120 MHz was actually split into separate uplink and downlink channels, and so a 20 MHz license typically included a 10 MHz uplink and a 10 MHz downlink channel, and is sometimes written as 2×10 MHz, or 10 MHz for short. In addition the UK and Italian auctions included some other, less valuable spectrum for what are called time division duplex (TDD) applications.

15. The Deutschmark was worth about 0.5 euros. It was replaced by the euro shortly after this auction ended.

16. Klemperer's (2004) account is a bit different. He indicated that T-Mobile and Mannesmann should have either dropped out two blocks earlier or kept going until one of the entrants dropped out. This would have been a sound pre-auction recommendation if T-Mobile and Mannesmann had both believed that the entrants were either weak or quite strong. As it was, each bidder faced a great deal of uncertainty about rivals' valuations, and as the two strongest incumbents, neither T-Mobile nor Mannesmann wanted to be the first to cede the third block, and possibly lose market share. Only when prices rose so high that a re-auction could be expected to result in lower prices for the unsold extra block did T-Mobile find it advantageous to reduce eligibility.

17. These four blocks were re-auctioned ten years later for a total of EUR 350 million, or less than 2.5 percent of the earlier auction price. Moreover Telefonica spent approximately EUR 1 billion on network buildout before deciding to walk away.

18. See Loxley and Salant (2004).

19. See Ausubel and Cramton (2010).

20. See http://www.puc.texas.gov/agency/rulesnlaws/subrules/electric/25.381/24492adt.pdf

21. See http://www.puc.texas.gov/agency/rulesnlaws/subrules/electric/25.381/24492adt.pdf.

22. See http://wireless.fcc.gov/auctions/default.htm?job=auction_summary id=66.

23. This type of rule is an application of the logic of the axiom of revealed preference (see Samuelson 1938). The CCA is discussed in more detail below.

24. See also Bulow, Levin, and Milgrom (2009) for a discussion of the bidding strategy that recognized the flaw in the activity rule. Bulow et al. did not include Cricket in their report. Salant (1997) also provides an example of how a bidding strategy can be adapted to take advantage of rivals' budget limitations and limited arbitrage opportunities in an auction.

25. See Bulow, Levin, and Milgrom (2009) for a discussion of Spectrumco's strategy. I advised Cricket.

26. See Salant (1997) for a more detailed discussion.

27. The Regulatory Authority managing the auction can try to set increments based on activity, as the FCC does a variable price increment rule, that is, one in which the size of the increment is based on bidding activity. See Cramton and Schwartz (2000) for a discussion.

28. See Loxley and Salant (2004) about the New Jersey BGS auction, which typically takes a few days. These auctions can be conducted much more quickly, and this would save both the bidders and the utilities running the auction money.

29. Variations of such rules have been tried in some recent combinatorial clock auctions, which are discussed in the next chapter.

Chapter 10

1. See Goeree and Holt (2010).

2. See http://wireless.fcc.gov/auctions/.

3. See Goeree and Holt (2010) for details.

4. See Porter et al. (2003).

5. Much of this section is from Day and Milgrom (2008) and Day and Milgrom (2010).

6. One of the 1,800 MHz blocks was 20 MHz, and the blocks were not all identical. However, Swisscom won better blocks than Sunrise.

7. For a discussion of stable and perfect equilibria see Van Damme (1991) or Kohlberg and Mertens (1986). Note that for one bidder to offer anything between 100 and 200 for all three lots, and zero for anything less is also a Nash equilibrium when its rivals both offer zero. However, if one of the rivals were to offer $x > 0$, any bid of less than $100 + x$ for the three lots is no longer a Nash equilibrium. Thus offers of between 100 and 200 for all three lots by one bidder, and zero by other bidders is not *stable.*

8. See Szentes and Rosenthal (2003) for an analysis of this type of auction with two bidders in a first-price sealed-bid auction. They find that there is an equilibrium in mixed strategies.

9. See Bulow, Levin, and Milgrom (2009) and Salant (1997).

10. The notion of the core is discussed in Gilles (1959). Also Edgeworth (1881) discussed a solution which is essentially the core in the context of a simple exchange economy.

11. See Ausubel and Milgrom (2002).

12. Also the ascending proxy auction is in the core.

13. In some CCAs bidders can at times bid on earlier-round packages whose points exceed eligibility when the relative price of the larger package has decreased. See Ausubel and Cramton (2011).

14. A revealed preference type activity rule is used to ensure consistency.

15. See Erdil and Klemperer (2010) for a discussion of alternative core adjustments.

16. There were some limitations (e.g., spectrum caps) that reduced this number somewhat.

17. The CCA differs from the ascending proxy auction of Ausubel and Milgrom (2002) in that there are no proxies in the CCA, and in the use of a second-price rather than a pay-as-bid rule.

18. Bakom had been provided examples of this possibility in meetings I had with them prior to the auction.

19. This example was suggested by Jon Levin and Andy Skrzypacz. See also Janssen and Karamychev (2013) for discussion of bidding strategy in CCAs.

20. Rosenthal and Szentes's (2003) paper was motivated by Rosenthal's work on a sealed-bid spectrum auction planned in the Netherlands in 1998.

21. If the bidders indicate an offer for one lot, two lots, or three lots, and the lots are identical, the analysis will differ, but there still won't be an equilibrium in pure strategies.

22. Ausubel and Milgrom (2002) have also noted the decision problem posed by budget constraints.

23. See Knapek and Wambach (2012).

24. See Ausubel and Milgrom (2002) for some discussion. See also Bichler, Shabalin, and Wolf (2011).

25. See ofcom.org.uk for the details.

26. See Bernheim and Whinston (1986). Also see Klemperer (2010) which discusses how similar ideas can be applied to other types of assets, such as financial ones with different ratings.

27. See Bernheim and Whinston (1986).

28. The following example, from Bernheim and Whinston (1986), illustrates this point. Suppose that there are two bidders, A and B, and two blocks, X and Y.

Block	Value for bidder A	E1 bid	E2 bid	E3 bid	E4 bid
X	6	5	0	7	0
Y	5	0	3	6	0
$Z = X \cup Y$	8	7	6	5	3
$\varnothing$	0	0	0	0	0

Block	Value for bidder A	E1 bid	E2 bid	E3 bid	E4 bid
X	5	0	2	7	0
Y	6	3	0	6	0
$Z = X \cup Y$	7	3	0	5	0
$\varnothing$	0	0	0	0	0

In this example, the strategies E1, E2, E3, and E4 are each an equilibrium. The revenues are, respectively, 7, 6, 5, and 3. The multiplicity arises from de facto coordination of the bids.

Chapter 11

1. See Kagel and Levin (2011) for a survey of behavioral and experimental analysis of auctions.

2. See Brown and Rosenthal (1990).

3. The Nextwave bankruptcy tied up a significant fraction of the available cellular frequency for over five years. See *FCC v. Nextwave Personal Communications Inc.* (01–653) 537 US 293 (2003) 254 F.3d 130.

References

Abreu, Dilip, David Pearce, and Ennio Stacchetti. 1990. Toward a theory of discounted repeated games with imperfect monitoring. *Econometrica* 58 (5): 1041–63.

Allaz, Blaise, and Jean-Luc Vila. 1993. Cournot competition, forward markets and efficiency. *Journal of Economic Theory* 59 (1): 1–16.

Ashenfelter, Orley. 1989. How auctions work for wine and art. *Journal of Economic Perspectives* 3 (3): 23–36.

Ashenfelter, Orley, and Kathryn Graddy. 2003. Auctions and the price of art. *Journal of Economic Literature* 41 (3): 763–87.

Ausubel, Lawrence M., and Peter Cramton. 2010. Virtual power plant auctions. *Utilities Policy* 18 (4): 201–208.

Ausubel, Lawrence M., and Peter Cramton. 2011.Activity rules for the combinatorial clock auction. Discussion paper. Department of Economics, University of Maryland.

Ausubel, Lawrence M., and Paul Milgrom. 2002. Ascending auctions with package bidding. *Frontiers of Theoretical Economics* 1 (1): 1–42.

Baldick, Ross, Ryan Grant, and Edward Kahn. 2004. Theory and application of linear supply function equilibrium in electricity markets. *Journal of Regulatory Economics* 25 (2): 143–67.

Bernhardt, Dan, and David Scoones. 1994. A note on sequential auctions. *American Economic Review* 84 (3): 653–57.

Bernheim, B. Douglas, and Michael D. Whinston. 1986. Menu auctions, resource allocation, and economic influence. *Quarterly Journal of Economics* 101 (1): 1–31.

Bichler, Martin, Pasha Shabalin, and Jurgen Wolf. 2011. Efficiency, auctioneer revenue, and bidding behavior in the combinatorial clock auction. Working paper. TU, Munchen.

Borenstein, Severin, B. James Bushnell, and Frank A Wolak. 2002. Measuring market inefficiencies in California's restructured wholesale electricity market. *American Economic Review* 92 (5): 1376–1405.

Brown, James N., and Robert W. Rosenthal. 1990. Testing the minimax hypothesis: A re-examination of O'Neill's game experiment. *Econometrica* 58 (5): 1065–81.

Brusco, Sandro, and Giuseppe Lopomo. 2002. Collusion via signaling in simultaneous ascending bid auctions with heterogeneous objects, with and without complementarities. *Review of Economic Studies* 69 (2): 407–36.

Bulow, Jeremy, Jonathan Levin, and Paul Milgrom. 2009. Winning play in spectrum auctions. Working paper. National Bureau of Economic Research.

Capen, Edward C., Robert V. Clapp, and William M. Campbell. 1971. Competitive bidding in high-risk situations. *Journal of Petroleum Technology* 23 (6): 641–53.

Cramton, Peter, and Jesse A. Schwartz. 2000. Collusive bidding: Lessons from the FCC spectrum auctions. *Journal of Regulatory Economics* 17 (3): 229–52.

Day, Robert, and Paul Milgrom. 2013. *Optimal Incentives in Core-Selecting Auctions: Handbook of Market Design*. New York: Oxford University Press.

Day, Robert, and Paul Milgrom. 2008. Core-selecting package auctions. *International Journal of Game Theory* 36 (3–4): 393–407.

Deltas, G., and G. Kosmopoulou. 2004. "Catalogue"vs "Order-of-sale" effects in sequential auctions: Theory and evidence from a rare book sale. *Economic Journal* 114 (492): 28–54.

Edgeworth, Francis Ysidro. 1881. *Mathematical Psychics: An Essay on the Application of Mathematics to the Moral Sciences.* London: CK Paul.

Erdil, Aytek, and Paul Klemperer. 2010. A new payment rule for core-selecting package auctions. *Journal of the European Economic Association* 8 (2–3): 537–47.

Fan, Jun Byoung, Heon Cuihong, and Elmar Wolfstetter. 2013. Optimal information disclosure in a license auction with downstream interaction. Working paper. Korea University.

Fisher, Ronald A. 1922. On the mathematical foundations of theoretical statistics. *Philosophical Transactions of the Royal Society of London. Series A, Containing Papers of a Mathematical or Physical Character* 222: 309–68.

Friedman, James W. 1985. Cooperative equilibria in finite horizon noncooperative supergames. *Journal of Economic Theory* 35 (2): 390–98.

Garber, Peter M. 1989. Tulipmania. *Journal of Political Economy* –97 (3): 535–60.

Gilbert, Richard J., and David M. G. Newbery. 1982. Preemptive patenting and the persistence of monopoly. *American Economic Review* 72 (3): 514–26.

Gilles, D. B. 1959. Solutions to general non-zero-sum games: Contributions to the *Theory of Games* vol. iv. *Annals of Math Studies* 40: 47–85.

Goeree, Jacob K., and Charles A. Holt. 2010. Hierarchical package bidding: A paper and pencil combinatorial auction. *Games and Economic Behavior* 70 (1): 146–69.

Green, Edward J., and Robert H. Porter. 1984. Noncooperative collusion under imperfect price information. *Econometrica* 52 (1): 87–100.

Green, Jerry, and Jean-Jacques Laffont. 1977. Characterization of satisfactory mechanisms for the revelation of preferences for public goods. *Econometrica* 45 (2): 427–38.

Green, Richard J., and David M. Newbery. 1992. Competition in the British electricity spot market. *Journal of Political Economy* 100 (5): 929–53.

Gribik, Paul R. 1995. Designing an auction for QFgeneration resources in California: What went wrong? *Electricity Journal* 8 (3): 14–23.

Grimm, Veronika, Frank Riedel, and Elmar Wolfstetter. 2003. Low price equilibrium in multi-unit auctions: The GSM spectrum auction in Germany. *International Journal of Industrial Organization* 21 (10): 1557–69.

Hendricks, Kenneth, and Robert H. Porter. 1988. An empirical study of an auction with asymmetric information. *American Economic Review* 78 (5): 865–83.

Hortacsu, Ali, and Steven L. Puller. 2008. Understanding strategic bidding in multi-unit auctions: a case study of the Texas electricity spot market. *Rand Journal of Economics* 39 (1): 86–114.

International Telecommunications Union. 2012. *Exploring the value and economic valuation of spectrum. Report*. New York: ITU.

Janssen, Vladimir, and Maarten Karamychev. 2013. Gaming in combinatorial clock auctions. Discussion paper 13-027/VII. Tinbergen Institute.

Jehiel, P., and B. Moldovanu. 2003. An economic perspective on auctions. *Economic Policy* 18 (36): 269–308.

Jones, C., F. Menezes, and F. Vella. 2004. Auction price anomalies: Evidence from wool Auctions in Australia. *Economic Record* 80 (250): 271–88.

Joskow, Paul, and Edward Kahn. 2002. A quantitative analysis of pricing behavior in California's wholesale electricity market during summer 2000. *Energy Journal (Cambridge, Mass.)* 23 (4): 392–94.

Kagel, John H., and Dan Levin. 2008. Auctions: A survey of experimental research, 1995–2008. In John H. Kagel and Alvin E. Roth, eds., *Handbook of Experimental Economics*, vol. 2. Princeton: Princeton University Press.

Kahn, Alfred E., Peter C. Cramton, Robert H. Porter, and Richard D. Tabors. 2001. Uniform pricing or pay-as-bid pricing: A dilemma for California and beyond. *Electrici–ty Journal* 14 (6): 70–79.

Klemperer, Paul. 2004. *Auctions: Theory and Practice.* Princeton: Princeton University Press.

Klemperer, Paul. 2010. The product-mix auction: A new auction design for differentiated goods. *Journal of the European Economic Association* 8 (2–3): 526–36.

Klemperer, Paul D., and Margaret A. Meyer. 1989. Supply function equilibria in oligopoly under uncertainty. *Econometrica* 57 (6): 1243–77.

Knapek, Stephan, and Achim Wambach. 2012. Strategic complexities in the combinatorial clock auction.

Kohlberg, Elon, and Jean-Francois Mertens. 1986. On the strategic stability of equilibria. *Econometrica* 54 (5): 1003–37.

Krishna, Vijay. 2009. *Auction Theory*. San Diego: Academic Press.

Kwerel, Evan R., and Gregory L. Rosston. 2000. An insiders' view of FCC spectrum auctions. *Journal of Regulatory Economics* 17 (3): 253–89.

Larson, Nathan, and David Salant. 2003. *Equilibrium in wholesale electricity markets*. Petten: Energy Research Centre of the Netherlands.

Levin, J., and A. Skrzypacz. 2013. Properties of the combinatorial clock auction. Working paper. Stanford University.

Loxley, Colin, and David Salant. 2004. Default service auctions. *Journal of Regulatory Economics* 26 (2): 201–29.

Marshall, Robert C., and Leslie M. Marx. 2007. Bidder collusion. *Journal of Economic Theory* 133 (1): 374–402.

McAdams, David. 2007. Adjustable supply in uniform price auctions: Noncommitment as a strategic tool. *Economics Letters* 95 (1): 48–53.

McAfee, R. Preston, and Daniel Vincent. 1993. The declining price anomaly. *Journal of Economic Theory* 60 (1): 191–212.

Mezzetti, C. 2011. Sequential auctions with informational externalities and aversion to price risk: Decreasing and increasing price sequences. *Economic Journal* 121 (555): 990–1016.

Milgrom, Paul. 1996. Procuring universal service: Putting auction theory to work. Lecture before the Royal Swedish Academy of Sciences, 9.

Milgrom, Paul. 2004. *Putting Auction Theory to Work*. New York: Cambridge University Press.

Milgrom, Paul R., and Robert J. Weber. 1982. A theory of auctions and competitive bidding. *Econometrica* 50 (5): 1089–1122.

Mueller, Milton. 1993. New Zealand's revolution in spectrum management. *Information Economics and Policy* 5 (2): 159–77.

Myerson, Roger B. 1981. Optimal auction design. *Mathematics of Operations Research* 6 (1): 58–73.

Paarsch, Harry J., and Han Hong. 2006. *An Introduction to the Structural Econometrics of Auction Data*. Cambridge: MIT Press.

Porter, David, Stephen Rassenti, Anil Roopnarine, and Vernon Smith. 2003. Combinatorial auction design. *Proceedings of the National Academy of Sciences of the United States of America* 100 (19): 11153–57.

Porter, Robert H. 1983. A study of cartel stability: The joint executive committee, 1880–1886. *Bell Journal of Economics* 14 (2): 301–14.

Riley, John G., and William F. Samuelson. 1981. Optimal auctions. *American Economic Review* 71 (3): 381–92.

Riordan, Michael H., and J. David Salant. 1994. Preemptive adoptions of an emerging technology. *Journal of Industrial Economics* 42 (3): 247–61.

Rothkopf, M. H., T. J. Teisberg, and E. P. Kahn. 1990. Why are Vickrey auctions rare? *Journal of Political Economy* 98 (1): 94.

Salant, David. 2000. Auctions and regulation: Reengineering of regulatory mechanisms. *Journal of Regulatory Economics* 17 (3): 195–204.

Salant, David. 2005. Multi-lot auctions. In Michael Crew, ed., *Obtaining the Best from Regulation and Competition*. Berlin: Springer, 41–64.

Salant, David J. 1997. Up in the air: GTE's experience in the MTA auction for personal communication services licenses. *Journal of Economics and Management Strategy* 6 (3): 549–72.

Samuelson, Paul A. 1938. A note on the pure theory of consumer's behaviour. *Economica* 5 (17): 61–71.

Selten, Reinhart. 1975. Reexamination of the perfectness concept for equilibrium points in extensive games. *International Journal of Game Theory* 4 (1): 25–55.

Serentschy, G. 2013. Analyse des Verlaufs der Multiband-Auktion 2013. Rundfunk und Telekom Regulierungs GMbH. Available at: https://www.rtr.at/en/pr/PI28102013TK/PK28102013TK_Praesentation.pdf.

Szentes, Balázs, and Robert W. Rosenthal. 2003. Three-object two-bidder simultaneous auctions: Chopsticks and tetrahedra. *Games and Economic Behavior* 44 (1): 114–33.

Tirole, Jean. 1988. *The Theory of Industrial Organization*. Cambridge: MIT Press.

Van Damme, Eric. 1991. *Stability and perfection of Nash equilibria*. New York: Springer.

Van den Berg, Gerard J., Jan C. van Ours, and Menno P. Pradhan. 2001. The declining price anomaly in Dutch rose auctions. *American Economic Review* 91 (4): 1055–62.

Varian, Hal R. 2007. Position auctions. *International Journal of Industrial Organization* 25 (6): 1163–78.

Vickers, John. 1986. The evolution of market structure when there is a sequence of innovations. *Journal of Industrial Economics* 35 (1): 1–12.

Vickrey, William. 1961. Counterspeculation, auctions, and competitive sealed tenders. *Journal of Finance* 16 (1): 8–37.

Vincent Lyk-Jensen, S., and O. Chanel. 2007. Retailers and consumers in sequential auctions of collectibles. *Canadian Journal of Economics. Revue Canadienne d'Economique* 40 (1): 278–295.

Walras, Leon, and William Jaffé. 1954. *Elements of Pure Economics; or, The Theory of Social Wealth*. London: American Economic Association/Royal Economic Society.

Wolak, Frank A. 2003. Measuring unilateral market power in wholesale electricity markets: The California market, 1998–2000. *American Economic Review* 93 (2): 425–30.

Wolfstetter, Elmar. 1996. Auctions: An introduction. *Journal of Economic Surveys* 10 (4): 367–420.

Index